D1444172

ETHICS, REPRODUCTION AND GENETIC CONTROL

ETHICS,
REPRODUCTION AND GENETIC CONTROL

Edited by RUTH F. CHADWICK

CROOM HELM
London • New York • Sydney

© 1987 Ruth F. Chadwick
Chapter 5 © 1987 R. M. Hare
Croom Helm Ltd, Provident House, Burrell Row,
Beckenham, Kent BR3 1AT
Croom Helm Australia, 44 – 50 Waterloo Road,
North Ryde, 2113, New South Wales

Published in the USA by Croom Helm
in association with Methuen, Inc.,
29 West 35th Street, New York, NY 10001

British Library Cataloguing in Publication Data
Desire and design: ethics, reproduction and
    genetic control.
    1. Human reproduction — Moral and ethical
    aspects
    I. Chadwick, Ruth F.
    176   QP251
    ISBN 0 – 7099 – 3472 – 6

Library of Congress Cataloging in Publication Data

ISBN 0 – 7099 – 3472 – 6

Printed and bound in Great Britain by Mackays of Chatham Ltd, Kent

Contents

Acknowledgements

I owe a great deal to the late Professor Sir Hans Krebs, who gave me valuable advice and suggestions, and encouraged me to learn Genetics.

I am very grateful to David Roberts, of the Department of Genetics, Oxford, for teaching me Genetics, to Jonathan Glover for supervising my B.Phil. and D.Phil. work on eugenics and genetic engineering between 1975 and 1979, and to my examiners Ted Honderich and Julie Jack.

I have also been helped by discussions with Professor W. F. Bodmer, Alasdair Houston, John Milbank, Professor Bob Williamson, my Medical Ethics students at S. Martin's College, Lancaster, audiences at various conferences, and, as is appropriate given the topic, my mother.

My thanks go also to Peter McNiven, of the John Rylands University Library of Manchester, and to W. J. M. Ricquier, of the National University of Singapore, for providing me with useful references and material.

Finally and most importantly I am grateful to those who have contributed their papers to this volume.

Jerome Lejeune's paper 'Test Tube Babies are Babies' is reprinted from *The question of in vitro fertilization* (Society for the Protection of Unborn Children, London, 1985) by permission of the author and the publisher.

'Marriage and the Family' is taken from *Personal origins: a report of the Working Party on Human Fertilisation and Embryology of the Board for Social Responsibility* (CIO Publishing, 1985) by permission of the Central Board of Finance of the Church of England.

Sir David Napley's paper, 'IVF and the Law' was delivered at the 1983 Mogul Conference (see below).

R. M. Hare's paper, '*In vitro* fertilisation and the Warnock Report' was delivered at the summer, 1985, meeting of the Hastings Center in Oxford, and the first half of it was delivered to a conference held in 1983 in Grosvenor House, London, convened by Mr. M. I. Mogul.

Robert L. Sinsheimer's paper, 'The Prospect of Designed Genetic Change', is reprinted from *Engineering and Science Magazine*, April 1969 (published at the California Institute of

Technology), by permission of the author and the publisher.

W. French Anderson's paper, 'Human Gene Therapy: Scientific and Ethical Considerations' is reprinted from the *Journal of Medicine and Philosophy*, vol. 10, no. 3, pp. 275–91 (copyright © 1985 by D. Reidel Publishing Company, Dordrecht, Holland), by permission of the author and the publisher.

C. K. Chan's paper, 'Eugenics on the Rise: A Report from Singapore', is reprinted from the *International Journal of Health Services* vol. 15, no. 4 (1985), pp. 707–12 (copyright © 1985, Baywood Publishing Co., Inc.) by permission of the author and the publisher.

Dharma Kumar's paper, 'Should One be Free to Choose the Sex of One's Child?' is reprinted from the *Journal of Applied Philosophy* vol. 2, no. 2 (1985), pp. 197–204, by permission of the author and the publisher.

Contributors

Jerome Lejeune, Professor of Fundamental Genetics, is Director of Research at CNRS, Université René Descartes, Paris.

BSR is the Board for Social Responsibility of the General Synod of the Church of England, Chairman the Rt. Rev. Hugh Montefiore, Bishop of Birmingham.

Sir David Napley is a distinguished lawyer and Senior Partner in Kingsley, Napley.

R. M. Hare, White's Professor of Moral Philosophy at the University of Oxford 1966–1983, is now Graduate Research Professor of Philosophy, University of Florida at Gainesville. His many publications include *The language of morals* (1952), *Freedom and reason* (1963) and *Moral thinking* (1981).

Robert L. Sinsheimer, biologist and author of more than 200 scientific papers, is now Chancellor of the University of California at Santa Cruz.

W. French Anderson, MD, is chief of the Laboratory of Molecular Hematology, National Heart, Lung, and Blood Institute, National Institutes of Health, Bethesda, Maryland, USA.

Chee Khoon Chan is a lecturer in the School of Social Science in the Universiti Sains Malaysia and is currently on study leave at the Harvard School of Public Health.

Dharma Kumar is a member of the Delhi School of Economics, Delhi, India.

Preface

Social ethics is not the province of philosophers alone. What philosophers can contribute is an examination of arguments, to see which are good and which are bad. R. M. Hare makes this point in this volume.

This collection brings together writings from different disciplines, to show various perspectives on the ethical issues that are now confronting us in the field of reproduction. Thus there are contributions from scientists, philosophers, social scientists, a lawyer, and the Church of England's Board for Social Responsibility.

It is important to know what is happening. Nothing actually follows from the facts as to what we ought to do, but facts may determine what our options are.

I have not spent time describing the various technologies in what has become known as the 'Reproduction Revolution' as in the title of the book by Peter Singer and Deane Wells. They have, I think, received sufficient publicity. However, where eugenics and genetic engineering are concerned, the possibilities may be less widely known, so they have been explained in a little more detail, both in the introductory chapter to Part II and by the papers in that section. It is instructive to read Robert Sinsheimer's forecast for the prospect of genetic change, together with the issues he raises, and compare that with contemporary accounts of what is feasible now. We might have thought what Sinsheimer described as the 'old' eugenics would be a thing of the past, but C. K. Chan's paper describes a recent eugenic programme in Singapore.

This paper, together with the one by Dharma Kumar, reminds us that we should not see issues of reproductive control purely from the perspective of modern western society, but that we need to be aware of the possibilities for, and viewpoints of, other societies.

Issues of IVF and genetic control have been dealt with in a single volume because I want to stress the connection between the two. The Warnock Report recommended that reproductive technology should not be used solely to help the infertile. (In the discussions that follow I use 'Warnock' to refer to the Report. When talking of Lady Warnock's own Introduction to the later edition I refer to her by that title.)

Another point that I particularly want to bring out is the way these developments will affect relations between and attitudes towards men and women and their roles and value in society.

The central argument, however, concerns to what extent we should be guided by the aim of trying to satisfy the desires both to have a child and to have a child of a certain quality. This involves an examination of the nature of these desires.

The view for which I want to argue is that we cannot simply be concerned with the desires of adults, but that we need to be sensitive to wider social consequences, such as what kind of problems we may be bequeathing to the children produced. In this context, it is argued, public policy should take a child-centred approach. I am assuming a view which does give some weight to the welfare of future children. Anyone who thinks we need not concern ourselves at all with the effects of our actions on as yet unborn generations will not agree with the views expressed here.

A full discussion of the issues of policy should also deal with questions of resource allocation, but this I have not had the space to do more than mention. It would require a volume in itself.

Ruth F. Chadwick
University College
Cardiff

Part I
Having Children

1

Having Children: Introduction

Ruth F. Chadwick

The phrase 'having children' is deceptively simple. On the face of it, it describes an activity that is practised universally and always has been. But the process is surrounded by an ever-increasing number of problems. Childbirth itself is a subject of controversy, as to how it should best be managed. To what extent the state should interfere between parents and children is another live issue.

For present purposes the important questions concern the appropriate response to infertility and the use of reproductive technology. These have received considerable attention of late, but some of the individual techniques have been considered to a greater degree than the underlying questions of value.

In looking at such value questions, it is sometimes difficult to know where to begin. Some people take the view that the burden of proof lies on those who would oppose new developments, to show why they should not be used. The presumption here is that individuals should be free to do as they choose unless there are good reasons to the contrary. Some of the papers in this volume do consider possible arguments against the new technologies. However, it seems important also to look at value positions underlying the arguments *in favour* of reproductive technology. For the debates in the literature have suggested that the choice to reproduce or not is a particularly important aspect of freedom. It will be helpful to examine why this is so, if it is. So we shall begin by looking at the question of whether there is a right to reproduce.

Is there a right to reproduce?

Article 12 of the European Convention on Human Rights states: 'Men and women of marriageable age have the right to marry and to found a family, according to the national laws governing the exercise of this right.' The wording of this is ambiguous. Does it imply that marriage is a necessary preliminary to the founding of a family? Whether it could support a right for unmarried people to have children is a matter of controversy, as is the question as to whether it could provide a justification of an entitlement to reproduce by artificial means. At the time the Convention was drafted, test tube babies were far from becoming a reality. The function of a right to reproduce was traditionally to prevent interference in individuals' lives by such methods as compulsory sterilisation. In other words, it was seen as a negative right with a corresponding duty of non-interference.

A distinction is commonly drawn between such negative rights and positive rights, which suggest that right-holders are entitled to have positive steps taken to enable them to exercise their rights. So a positive right to reproduce could indicate the provision of reproductive technology.

Not too long ago such a claim would scarcely have been intelligible, where a positive right in this context would have involved the provision not only of a sexual partner, but also of a fertile one.

But it may be argued that the situation has changed. Where means are available, a right previously negative in scope can become a positive one. Charles Fried argues in another context that if there are any positive rights at all, there is only one, viz. a right to a fair share of society's resources.[1] Where those resources include reproductive technology, perhaps everyone has a right of access to them. Thus Sheila Maclean argues that once means are available they may not be distributed in a discriminatory way:

> whilst the state may have no general duty to facilitate reproduction through technology or the supply of a partner, once facilities are provided — for example through *in vitro* fertilization and surrogacy programmes — to deny access on grounds of sexuality is to infringe the right on a discriminatory basis.[2]

It may be argued that there is no real distinction between negative and positive rights. For if means are available and the state denies access to them that may be construed as an interference with freedom just as surely as compulsory sterilisation. It might seem odd to take the view that there is no difference between compulsory sterilisation and refusal to offer *in vitro* fertilisation (IVF), but this may be because of factors involved in the means of violation of the right, as opposed to the principle of violation itself. We need however to look at what is the basis, if any, for the right, at its justification, to see if it provides an argument for a positive right.

As John Harris has pointed out,[3] rights can be treated either as the starting points or as the conclusions of argument. Those who use them in the first sort of way may regard them as natural, or self-evident.

A natural right to reproduce

It is not clear how we are to understand 'natural' rights; whether they are anything more than 'nonsense on stilts', as Jeremy Bentham described them.[4] Perhaps the suggestion is that the right to reproduce is self-evident. But it is certainly not self-evident to many people, e.g. to those who see a child as a gift from God, as a blessing on a marriage, but certainly not as something to which couples or individuals have a right.

Further, there are clearly difficulties in seeing the right to reproduce as a 'natural' right where what is under discussion is the provision of artificial reproductive technology to the infertile. There seems to be something very odd about saying that persons have a natural right to reproduce by artificial means.

However, Peter Singer and Deane Wells have argued that *natural law* could perhaps provide an argument for the use of reproductive technology. In discussing *in vitro* fertilisation they point out that as natural law holds that the purpose of sexual intercourse is the production of children, and thus rules out contraception, it ought also to hold that to refuse to assist infertile couples to reproduce is to connive at a subversion of the purpose of the sex act.[5]

A difficulty with this is that natural law does not hold that it is desirable that *every* individual should reproduce — otherwise it would have been unable not only to condone but also to approve positively of chastity. It stresses the reproduction of the species rather than that of the individual.

S. L. Floyd and D. Pomerantz have considered whether there is a natural right to reproduce, and have not come up with any very encouraging conclusions for those who would wish to take this view.[6] They consider that it might be seen as an aspect of self-determination or as an aspect of the right to do with one's body as one wishes, but conclude that neither of these avenues is satisfactory for two main reasons. First, having a child involves other people, not only oneself, and secondly, it produces a new person who is not consulted. So it is not simply a question of *self-*determination or of using one's own body as one wishes. Even if one ignores the difficulty about the child's wishes and restricts consideration to those presently existing, it is hard to see how a right to self-determination can, without more, provide an argument for the co-operation of others in providing either sexual intercourse or technology. So the prospects for a positive right to reproduce by artificial means look grim on this view.

The right to privacy

An alternative view is that a right to reproduce can be seen as an aspect of the right to privacy.

In the US case of *Eisenstadt* v. *Baird*, Justice Brennan said:

> If the right of privacy means anything, it is the right of the individual, married or single, to be free from unwarranted governmental intrusion into matters so fundamentally affecting a person as the decision whether to bear or beget a child.[7]

To discuss whether there is a right to privacy is not our concern. The important question is whether or not reproduction is a private matter. There may be something in the suggestion that reproduction should be something for which society as a whole assumes responsibility, even at the begetting stage. This need not involve Plato's mating festivals or the type of ticket system for sexual intercourse envisaged in Zamyatin's *We*. It *might* involve public decision-making about resources, e.g. whether society can afford to waste the resource of large numbers of aborted foetuses on the one hand while on the other hand spending large amounts of money on providing *in vitro* fertilisation.

The most familiar arguments for social responsibility take place concerning the *rearing* of children. One advocate of shared responsibility is Penny Perrick, writing in *The Times*:

> Far from urging less government interference I would like to

see the appointment of a Minister for Children. And please let it be . . . somebody who insists that the basic rudiments of parenthood are taught as part of the school curriculum.[8]

Similarly, a case can be made for public concern with rearing by the argument that if society wants children then there should be public provision, in the form of funding out of the public purse for such facilities as day nurseries. Should public concern apply only to rearing and not to begetting?

Lady Warnock, in her Introduction to the later edition of the report, *A question of life*, invokes the distinction between what is public and what is private in explaining why the Committee decided not to recommend that the law should try to ban private surrogacy arrangements, whereas recommendations were made that commercial agencies should be prohibited. She suggests that 'a law against agencies would not be intrusive into the private lives of those who were actually engaged in setting up a family', whereas a law against all surrogacy would be.[9] The distinction between what is public and what is private is also invoked to explain why embryo research cannot be seen as a matter of conscience.

In her discussion of this topic Lady Warnock attempts to relate the discussion of the public and private to the Hart-Devlin debate on the law and homosexuality in the late 1950s and early 1960s.[10]

She claims to reject Devlin's notion of a moral consensus within society which must be reflected in the law. Even if there could be such a thing on some issues, she says, there could not be one on such new and uncertain questions. Thus it will not do to say that the law should rule out a practice such as surrogacy on the grounds that there is a general consensus that it is morally wrong. Rather, she says that we should look at Hart's answer to such problems. Hart's position was that we must ask two questions, first whether a practice is harmful, and secondly, whether the infringement of liberty that would be involved in prohibiting it would be in itself harmful. Lady Warnock's view is that surrogacy may well be wrong but to prevent private surrogacy would be unduly intrusive.

On the other hand, placing legal restrictions on commercial surrogacy and on embryo experiments would not be unduly restrictive. These are public matters.

How then does she draw the distinction between the public and the private? She says: 'The reason for this . . . distinction . . . between what might be thought a private matter and one which was *necessarily* public was somewhat obscure.'[11]

Are there any clues which might make the reasoning less obscure? Let us consider her remarks concerning embryo research.

One criterion she mentions is that embryo research is publicly financed. But this, while sufficient to bring it into the public arena, is not a necessary condition for her purposes, for she says that even if embryo research were privately financed it should be controlled by law. This is inconsistent with a wholehearted application of the finance criterion.

Interestingly enough, though, having earlier rejected Devlin's idea of a moral consensus, she falls back on this herself when talking about why embryo experiments should be controlled. She says that there is a 'public and widely shared sentiment' that embryo experiments are wrong.[12]

Is Lady Warnock right to say that there is a consensus on the issue of embryo experiments? It seems clear that she is not. One has only to think of the heated debates in Parliament concerning the Enoch Powell Bill. It is clear also that there are those willing to argue in favour of embryo experiments on the grounds that they will benefit humankind as a whole.[13]

Lady Warnock's discussion has attempted to draw a distinction between private and commercial surrogacy, and between some reproduction issues and the issue of embryo research. But there is no discussion of whether there is a right to provision of IVF which can be seen as an aspect of the right to privacy. It is apparent, however, from what is said elsewhere in the report (e.g. that services should be limited to heterosexual couples),[14] that some public criteria are implied.

Whatever these criteria are, it seems clear that once services are provided, reproduction is taken out of the private sphere and into the public. It is difficult to see IVF as a private matter. This is partly because of the finance criterion. If the resources are publicly financed, then the public may reasonably be thought to have an interest in their proper allocation.

It might also be added that this criterion (as Lady Warnock suggested in connection with embryo experiments) is not the whole story. This may be seen as a public concern, not however because of some Devlin-type moral consensus within society, but because these technologies may change for the whole of society the way in which reproduction is regarded, and thus the way we think of ourselves.

Justice Brennan's statement quoted above has special relevance where an invasion such as compulsory sterilisation is envisaged, but it is difficult to see the right to privacy as providing a basis for a positive right to reproduce.

Assigning a right to reproduce

Perhaps the right to reproduce could be seen as the *conclusion* of an argument. The right in this sort of case would function very differently from in the former case. For what is argued is that there are moral arguments supporting the view that it is desirable for people to have children, which outweigh any arguments suggesting that they ought not to be allowed to do so. In other words, having looked at the moral arguments, the conclusion is that the persons concerned ought to have the freedom to have a child. Thus a right, negative or positive, could be *assigned* on the basis of a utilitarian argument. This is totally different from saying that the right is natural or self-evident.

On this sort of view it might seem that rights can be more easily overridden by other considerations which would better promote utility. In which case, then, some would suggest that the language of rights serves no purpose. Jonathan Glover, for example, argues that if rights are not absolute they do no useful work.[15]

But let us look at how the argument might go.

An argument which is very frequently encountered in the literature is that based on the desire for a child. It is said that only those who have experienced it can know just how all-consuming it is.

The desire for a child

The first question to be borne in mind is: what is the content of this desire? It is helpful, when speaking of reproduction issues, to keep in mind the distinction between the concepts of begetting, rearing, and bearing. In some discussions there is talk simply of 'the desire for a child', without specifying what exactly is desired. At other times it is assumed that one or another of the above is meant, without showing good reasons for such an assumption. For example, Warnock says: 'In addition to social pressures to have children, there is, for many, a powerful urge to perpetuate their genes through a new generation. This desire cannot be assuaged by adoption.'[16] But it is not justifiable to make this assumption about the content of the desire without investigation, and the particular desire which is in question can make a very large difference to the conclusions that may be reached. For example, however important we may feel it is to enable people to satisfy a desire to pass on their genes, this could not, *without more*, provide an argument for male pregnancies (one of the latest suggestions for the application of medical expertise).[17]

Let us look more closely at the differences involved.

The desire to rear

In principle it is possible for this desire to be satisfied for everyone without recourse to artificial reproductive technology. For if individuals are unable for some reason to have their own children, they may rear those of other people. It may of course be impossible in practice owing to a shortage of babies for adoption.

Given such a shortage, this desire may be important in the surrogacy debate. In the type of case where the carrying mother not only provides her womb but also her egg, the commissioning mother is enabled only to rear a child but not to beget or to bear.

It is beyond question that the desire to rear a child can be and often is a very strong one. It is this that is at issue in custody disputes.

The desire to bear

Traditionally this has been a desire that only (but not all) women could satisfy. Even reproductive technology however, as presently envisaged, cannot make this possible for every woman. For some it may involve a serious risk to their health to bear a child.

Recently there has been a suggestion that there are some males who would like to become pregnant. The demand is said to come partly from transsexuals and partly from men who would like to bear children for their wives. Whatever the reason (they may simply want to have a new experience) it is important to consider this desire as another aspect of wanting to have a child.

To some it may seem that in any case a man cannot bear a child in the same sense that a woman can. We need to distinguish between carrying, and giving birth. It is clear from the recent Wendy Savage case that it is important for some *women* not only to carry but also to give birth. But for others the birth experience has been so technologised that it has come to seem a terrifying and alien experience. As Germaine Greer points out:

Yet women continue to want to bear children. They may say that they want to experience childbirth. What they can mean by that when there is no telling whether they will be allowed to experience it, given the aggressiveness of childbirth management, is not at all clear. Far too many women have no experience of birth at all, but simply of anaesthesia . . . Women who want the experience of childbirth are in the curious position of desiring the unknown.[18]

The desire to beget

The term 'beget' is used more commonly of men than of women. But understanding this term in the sense of passing on genes, it can be used sensibly for both parents.

This is the desire that Warnock saw as the significant one. But as we have already seen, we cannot assume that this is the only desire in question, otherwise all methods which involve a donor would not be satisfying the desires of at least one party of a donee couple (assuming a case where there is a couple both of whom have a desire for a child).

One interesting result of seeing this as the central desire, however, is that in principle, if it is thought to provide an argument for enabling a person to reproduce, it seems also to provide an argument for cloning.[19] Is there a significant difference between wanting to pass on *half* my genes in sexual reproduction, and wanting to pass on *all* my genes, as in cloning? It is not clear what the difference is.

In addition to these possible aspects of the desire to reproduce, there are other factors that might be involved.

The desire to have a child with someone

It might be argued that the above distinctions are missing the point. Surely the central case (the argument might go) of having a child is wanting to have a child with a particular person, as an expression of that relationship. The excitement comes from seeing the result as a mixture of the two of you. This seems to have been completely overlooked in the debate on reproductive technology. Of the new methods, *in vitro* fertilisation provides a way to satisfy this particular desire, in the sort of case where technology is used simply in order to assist a man's sperm in the fertilisation of his partner's egg. But any method that involves gamete donation does not.

But much of the discussion of these issues in the literature

11

sees the desire for a child as a desire for something for *oneself*. Much is made of enabling a person to have a child of his or her own. Thus one argument in the Warnock Report supported surrogacy as it provided the only way for some men to have a child of their own.[20] Perhaps the child is seen as an extension of the self. In wanting to pass on one's own genes, maybe one is making a bid for a kind of immortality. Similarly one might want to bear a child for oneself, as an experience one wants to have, like wanting to have the experience of looking at the Grand Canyon.

Socially induced desires

It is envisaged by some that the desire for a child may be an aspect of a desire to appear as a 'normal' family. Thus a letter in *NACK*, the *Quarterly Journal of the National Association for the Childless*, points to the problems caused by 'the lack of sons-in-law and daughters-in-law, no grandchildren, the silver wedding with no family, the passing round of photos and stories of grandchildren'.[21] The social pressure is recognised by Warnock:

> Family and friends often expect a couple to start a family, and express their expectations, either openly or by implication . . . Parents likewise feel their identity in society enhanced and confirmed by their role in the family unit.[22]

The desire for an heir

In societies where property and inheritance are valued, the desire for an heir plays a large part in the desire for a child. This provides a perfect example of the extent to which social factors influence the desire to reproduce. Women have been divorced for failing to provide an heir.

The history of the concept of legitimacy is important here. The desire for an heir has traditionally involved not only a desire for a child but for a legitimate child. This brings out the point that in societies where legitimacy is an important factor the desire for a child is not nearly so central as the desire to have one conceived in the appropriate way. Passing on one's property is thus held to be more important than passing on one's genes.

The significance of this point lies not in what is said about legitimacy itself but in the light it throws on the nature of the desire for children. To what extent is it a socially produced desire?

The desire for a child: natural or artificial?

Peter Singer and Deane Wells, in *The reproduction revolution*, take the view that 'the desire for children is . . . something very basic and cannot be overcome without great difficulty, if at all. There are obvious evolutionary reasons why this should be so.'[23] On the other hand, Fanny Lines, in a letter to the *Guardian*, writes:

> You talk of 'infertility blighting the happiness of couples'. Others speak of the desperation of childlessness. These are emotive, strong words. The fundamental problem is the desire for a child. Why? These people have been pressured by our culture, conditioned by a society that values inheritance, woman's mothering role and man's role as provider.[24]

So, is the desire biological or is it social? Singer and Wells stress the evolutionary reasons why it might be thought that the desire is basic to human beings. But if the desire for children is so basic, why has it evolved in connection with a very strong desire for sex, to back it up?

According to Plato, the desire for children is natural as a desire for immortality:

> there is a sense in which nature has not only somehow endowed the human race with a degree of immortality, but also implanted in us all a longing to achieve it, which we express in every way we can. One expression of that longing is the desire for fame and the wish not to lie nameless in the grave. Thus mankind is by nature a companion of eternity, and is linked to it, and will be linked to it, for ever. Mankind is immortal because it always leaves later generations behind to preserve its unity and identity for all time: it gets its share of immortality by means of procreation. It is never a holy thing to deny oneself this prize, and he who neglects to take a wife and children does precisely that.[25]

The fact that Plato in the *Laws*, however, felt obliged to suggest penalties for non-participation in this enterprise,[26] suggests that it may not be as natural as it is cracked up to be by him.

There is also plenty of evidence to suggest that the desire for children is affected by social circumstances.[27] For example, family size has been correlated with social status, and with social

13

circumstances such as provision for looking after the elderly. Then we have also to consider the fact of large-scale abortion — what does this say about the desire for children? Germaine Greer has argued that it, among other things, shows that modern western society is profoundly hostile to children.[28]

To say that there is a very large element of social determination in the desire that people have for children is not to deny that there may be some biological element in it; it is simply to point out that we cannot assume that the desire for a child is a very basic fundamental one natural to human beings and that that is the end of the story.

It is important also to bear in mind that even if the desire to have children is to a large extent socially induced, that need not imply that it is any less *strong* than it would be were it an integral part of human nature. A socially induced desire may be just as strongly felt. This will be important if we find that the strength of desire as opposed to its origin is an important consideration.

It is important to make these distinctions, as the way in which we describe the desire can influence the way in which we think about it. Paul Ramsey has made the point that the very fact that we use the term 'reproduction' rather than e.g. procreation is significant: 'a significant move toward *in vitro* fertilization and all the rest was made when first we began to use a manufacturing term — ''reproduction'' — for procreation'.[29] Just as it may be important whether we speak of reproduction instead of procreation, just so it may be important whether we speak of begetting or rearing.

With our distinctions made, we can go on to assess whether we should assign a right to reproduce on the basis of a desire to do so.

The argument from desire

We have seen that the desire for a child may be one of a number of different desires, or perhaps a combination of them. What we have to consider now, however, is whether the fact that such a desire exists constitutes a good argument for the use of reproductive technology.

On the face of it, it does not follow from the fact that someone wants something, that s(he) ought to have it. This is clear not only from reflection on experience but from the gap between what is the case and what ought to be the case that was pointed out by Hume (according to the standard interpretation of Hume's famous passage). Further argument is required to show that the desire necessitates action, rather than a response of 'Too bad'.

One way in which the desire argument could be backed up would be by a utilitarian framework. For obviously within an ethical perspective which advocates desire satisfaction in general, such as preference utilitarianism, the presence of this desire will provide prima facie grounds for satisfying it. It would, of course, have to be weighed against the satisfaction of other desires claiming consideration, but those who use the desire argument seem to accept that it is a very strong one. Thus Warnock says: 'For those who long for children, the realisation that they are unable to found a family can be shattering. It can disrupt their picture of the whole of the rest of their future lives.'[30] Similarly, Singer and Wells make the point that 'the need for IVF can be established by the strength of the desire for children'.[31]

Singer and Wells are, indeed, taking a utilitarian approach. But Warnock in the Foreword rules out such an approach as being incapable of providing answers to the type of question under consideration.

A strict utilitarian would suppose that, given certain procedures, it would be possible to calculate their benefits and costs. . . . However, even if such a calculation were possible, it could not provide a final or verifiable answer to the question whether it is *right* that such procedures should be carried out.[32]

It is important to be clear about the underlying reasoning, because the argument is often presented as if it follows from the fact of the desire that there is a need for IVF, without more. But this is obviously not the case. Thus Fanny Lines writes:

there are millions of childless people who do not 'suffer' but get on with living and make valuable contributions to society. Is not the answer a more general acceptance of all sorts of patterns of living of which the nuclear family is only one?[33]

Fanny Lines is making the point that the fact of a desire does not necessitate the conclusion that we ought to seek its satisfaction: on the contrary, those who have the desire may find that they can achieve just as much fulfilment by looking for alternatives.

But proponents of the desire argument who, like Warnock, do

not want to be utilitarians, may argue that of course we do not think that the simple fact of a desire necessitates its satisfaction. But there is something special about the *content* of the desire. Perhaps it is the fact that the desire is a desire *for a child* that wins it such support. Then of course the basis of the argument has shifted from the fact of desire to its concern with children. Perhaps the idea is that children are a good thing to have around; why is this?

Of course, there have been societies, and still are, where *governments* want to try to encourage reproduction (especially by certain sections of the population, as we shall see later), but if this is the case, it might be unwise to concentrate on this particular method (viz. the use of artificial means). If government is serious about wanting to encourage people to have more children, to avoid a shortage of citizens, or soldiers, more care facilities might have greater practical results than technology.

However, perhaps this is not the point. It may be that there is a very deep-seated belief among people generally, that children are a good thing, a blessing. Perhaps it is felt that the hope and future of the species lies in its children. For we must remember that in this argument we are no longer thinking in terms of the desires of individuals.

There are well-known difficulties in explaining why we should value the creation of new people,[34] but even if for some reason children are felt to be a good thing, so that the more the merrier, it would not necessarily indicate reproductive technology. We could cut down on abortions and thereby ensure that more children were produced. But that would involve overriding the desires of those who want abortions, which brings us back to the argument from desire.

Without some backing such as utilitarianism, the argument from desire is a non-starter. But a right to reproduce could be assigned on the basis of a utilitarian argument. How adequate would this be?

Objections to the argument from desire

Some object to the argument from desire on the grounds that it overlooks important considerations, or that it is inappropriate in this context.

Warnock makes the point that 'Moral questions are, by definition, questions that involve not only a calculation of consequences,

but also strong sentiments with regard to the nature of the proposed activities themselves.'[35] R. M. Hare's paper in this volume,
however, argues that sentiments cannot be the starting points of
moral arguments. If people's sentiments are outraged then that is
an important consequence to consider, but the sentiments themselves are not conclusive as to what is right or wrong.

Desires, needs, and medicine

Some would draw a sharp distinction between satisfying desires
and serving medical needs. Paul Ramsey, for example, points to

> a correlation . . . between the devotion of medical science to
> manufacturing the child as a 'product' . . . and the devotion
> of medical science to the treatment of human desires . . . The
> important line lies between doctoring desires . . . and seeking
> to correct a medical condition if it is possible to do so. Across
> this line, medical science has made a step . . . in attempting
> to produce a baby without first actually curing a woman's
> infertility.[36]

So is infertility a condition warranting medical treatment?

Leon Kass argues that infertility is not a *disease*. Although the
criteria for what counts as disease are by no means uncontroversial, Kass makes clear what his are when speaking of
infertility:

> It is not life threatening or crippling, nor does it lead to
> detectable bodily damage. To consider it a disease leads to a
> focus on an individual; yet infertility is a condition that is
> located in a marriage, in a union of two individuals. If it is
> any kind of disease, it is a 'social disease'.[37]

There are two ways of arguing against the point of view repre-
sented by Kass and Ramsey. The first is to argue that in fact these
people are ill or that infertility is a disease. The second is to argue
that it is irrelevant whether or not it is a disease.

Warnock adopts a mixture of the two, arguing both that the
medical system does not confine itself to treating needs anyway,
and that infertility is a 'malfunction'. The Committee does not go
as far as calling it a disease.

an inability to have children is a malfunction and should be considered in exactly the same way as any other. Furthermore infertility may be the result of some disorder which in itself needs treatment for the benefit of the patient's health . . . In addition the psychological distress that may be caused by infertility in those who want children may precipitate a mental disorder warranting treatment. It is, in our view, better to treat the primary cause of such distress than to alleviate the symptoms. In summary, we conclude that infertility is a condition meriting treatment.[38]

There are three different reasons given here for why infertility might be thought to fall into the 'medical' bracket. We have, however, been sufficiently warned of the dangers of the 'medicalisation' of more and more areas of human life,[39] such as bereavement. In order to avoid this a clear characterisation of what is a medical matter seems an attractive idea. Unfortunately, it is not easy to provide. What falls within the proper scope of medicine will be partly a matter of evaluation, just as our understanding of what counts as disease has been.[40] So it is not clear how useful it is to attempt to draw a line between what is medical and what is non-medical, and between a desire and a need.

For example, if we decided that medicine ought to be concerned only with needs, and we had a fairly strict interpretation of needs in this context, in terms of e.g. physical suffering, we might find that all preventive medicine and doctors' advice on contraception would be ruled out. After all, people who request contraceptive advice or who go for screening for cervical cancer are not, typically, experiencing physical suffering. On the other hand, it might be possible to construe all claims for medical help in terms of a desire, e.g. the desire to be well.

Whether we talk in terms of needs or desires, there is a perennial problem in that resources are finite and demand is infinite. So decisions will have to be made as to which demands to respond to, *and in what way.*

For example, the point Warnock makes about psychological attitudes is interesting. If people are having psychological problems about something in their lives, Warnock says that one should try to remove the cause, i.e. infertility, rather than alleviating symptoms by e.g. helping them to change their attitudes. But is it obvious what the cause is? Besides, elsewhere in the report Warnock has said that many of the problems about AID (artificial insemination by donor) are caused by an unwillingness to be open about it. We find the suggestion that 'a change in attitude towards male infertility is . . . required'.[41]

So the objection about the proper scope of medicine does not necessarily show that medicine should not concern itself with desires, but it does bring out the point that there is more than one possible response to desires, and this is an important contribution to the debate.

The argument from desert

One of Kant's objections to giving happiness a central place in morality was that some people just do not deserve to be happy.[42] Similarly it could be argued that those who have become infertile as a result of promiscuous behaviour should not be entitled to society's medical resources to remedy this. Infertility may be caused by sexually transmitted disease. Is this relevant?

First of all this argument involves the assumption that these people have 'brought it on themselves'. But even if that were true, those who might want to use this argument must acknowledge that it is analogous to the argument that people who have lung cancer as a result of smoking should not receive treatment for their condition. Indeed, there are some who take this view, but it is a very hard one to maintain consistently. Who can say she never takes any risks with her own health? Mountaineers, jockeys, drinkers, meat eaters — should they all be denied treatment if their problem can be traced to their chosen lifestyle? We all take risks every time we get into a car. This is not to deny, of course, that it might be advisable to discourage people from promiscuity. That is a different point, and one that is receiving considerable publicity in the current AIDS scare.

Limits

Warnock says 'There must be *some* barriers that are not to be crossed, *some* limits fixed, beyond which people must not be allowed to go'.[43] R. M. Hare's paper in this collection argues that there are no fixed rules when it comes to dealing with these novel problems. He suggests that it is not clear that rules which have gained fairly solid bases over time, such as the prohibition on adultery, can satisfactorily be extended to the new technologies. Even the Warnock Committee does not think the limits already exist. They speak of 'setting up' the barriers.[44] But they set them up in accordance with their feelings rather than in accordance with moral argument. It has to be admitted nevertheless that the barriers that are set up are relatively few, such as a time limit on embryo experiments.

The argument is, however, that there should be some limitations which cannot be satisfactorily provided by the argument from desire. Let us look at some possibilities.

Embryos

Some take the view, as Enoch Powell has done, that the new technologies have implications that are unpalatable in the lack of respect they show for human embryos. Others think that we should not be over-concerned with what happens to what is in reality little more than a collection of cells, particularly since we are willing to countenance such contraceptive practices as the intrauterine device.

Three views on embryos are put forward in this volume. Jerome Lejeune emphasises the viability of the early human embryo in an attempt to show that it should be regarded as a baby. Sir David Napley suggests that it should not amount to abortion to terminate the life of an embryo. R. M. Hare supports Warnock's conclusion that no answer can be given to the question of whether the embryo is a human being or not. Rather, we must decide what is the correct way to treat the embryo. He criticises, however, the way in which Warnock handles the question, and wants to replace it with a consequentialist approach. So, not only do we have to consider potential harms to these embryos, whatever status they may have, but these harms must be balanced against other harms and benefits. So on his view a utilitarian approach can adequately handle the question of embryos.

Children's rights

Some of the new technologies will produce children who do not know who one or both of their parents are. This may be thought to violate a right to know one's genetic history.

This does not really provide a problem for the utilitarian approach. For it will be important to consider the evidence from now adult AID children that the lack of such knowledge can adversely affect one's well-being.[45] This is because there can be a very strong desire to know. It is the stuff of which soap operas are made. A utilitarian may very well want to assign a right to know, and indeed it may be very difficult to argue for a right on any other grounds, as we found in the case of the right to reproduce.

An objection to the supposed difficulty about lack of knowledge is that in fact there are already many children conceived in the normal way who are blissfully unaware of who their real father is, so why should we worry about this with regard to babies produced by donation?[46]

This objection is confused. There are two choices when a child is produced by donation: either to tell her the truth, or to keep it a secret. If she is told the truth then she must be compared with children conceived normally who have also found out the truth, not with those who are blissfully unaware.

The keeping of a secret introduces further problems which are not specific to the fact of uncertainty of parenthood. Snowden and Mitchell have argued that it can have serious consequences for relationships, not only in the immediate family but also with aunts and uncles and grandparents. Relationships may be damaged by lack of openness.[47]

Of course, one possibility would be to recommend that all children born by donation should be given the legal right to know who their genetic parents are, as is the case with adopted children. But this was not a course taken by Warnock. Warnock advocated that donor anonymity should be preserved but that children born by e.g. AID should have the right to know basic facts about the donor's ethnic origin and genetic health.[48]

This may not be enough to counteract the unhappiness caused by lack of knowledge. At this stage, however, the important point for us to note is that the argument for children's rights is not a problem for a utilitarian approach which gives weight to the desires of children who will be produced by artificial reproduction.[49]

The family

One obvious possible casualty of the reproduction revolution is the institution of the family. Thus Jerome Lejeune's contribution to this collection envisages a time when the use of the word 'mother' will be an obscenity, as in *Brave new world*. But not all science fiction writers see it this way. In *Woman on the edge of time*, Marge Piercy depicts a society in which reproduction has been technologised but to act as mother to one of these children is viewed with great respect.

But to some it certainly seems a threatening possibility that the institution of the family may be undermined by the new reproductive technologies. Thus the Board for Social Responsibility says of surrogate motherhood: 'We are in accord, then, that in surrogate motherhood the Christian institution of the family is fundamentally endangered, and thus that it cannot be morally acceptable as a practice for Christians.'

The word 'family' is notoriously vague. The debate continues over the existence and importance, historically, of the 'nuclear' as opposed to the 'extended' family, and over whether the family depends on marriage.[50] For present purposes I shall assume that 'family' means the nuclear family based on husband and wife as father and mother.

There are two questions here: first, will the family be endangered, and secondly, why should that matter?

Singer and Wells have taken the view that childlessness itself constitutes a threat to the family or at least to marriage.[51] This may be true of particular families, but on the other hand the family as an institution is thought to be threatened by the introduction of third parties, as donors (albeit anonymously), into the nuclear family, and by the possibility that single persons or homosexual couples might decide to avail themselves of reproductive technology to have children. The more this happens, the less strong will be the nuclear family's claim to be the basic social unit.

Why should it matter if the family is endangered? It may be that those who wish to preserve it take the view that it has been instituted by God, or that the family represents the most satisfying way for human beings to live. These two claims may, though need not be, combined in the view that the family has been instituted by God precisely *because* it embodies the most satisfying way of life for human beings. Alternatively it might be argued, as it has been by Paul Johnson, that the family performs necessary social functions, and that its breakdown can be held responsible for such social problems as the inner-city riots of recent years.[52]

It may be considerations of this sort that led Sir David Napley to recommend that fertilisation outside marriage should be a criminal offence 'in the interests of society'.

The contribution from the Church of England's Board for Social Responsibility takes the view that 'the relief of childlessness should be undertaken only within marriage and as a service to marriage'. The view advanced is that children are one of the proper goods of marriage and are a good that is peculiar to marriage. It is within 'the context of family life in which the Creator wills young human beings to flourish'.

· The Board for Social Responsibility talks in terms of ends or purposes of marriage laid down by God. It is not necessary, however, to accept this perspective in order to agree that the provision of reproductive technology should be limited to married couples. Thus one could put forward an argument on utilitarian grounds that it is for the best that only married couples have children, because only in that relationship is there the kind of commitment necessary for the proper bringing up of children if secure and happy children are to be produced. R. M. Hare takes the view that it is the most valuable institution for that purpose.

Unmarried heterosexual couples

Some, however, while not wanting to go this far, do suggest that at least it should be limited to heterosexual couples. Thus Warnock says:

> we believe that as a general rule it is better for children to be born into a two-parent family, with both father and mother, although we recognise that it is impossible to predict with certainty how lasting such a relationship will be.[53]

Warnock introduces an element which is not just a point about children's flourishing, or lack of it, but about the role played by the family in socialisation. Warnock argues (a point also made by the Board for Social Responsibility) that there is a social aspect to bringing up children, that the family is the place where social behaviour is learned. Here we have a statement not about the security of the relationship, but about the presence of both a maternal and paternal role as the ideal, for children to develop.

It has long been recognised that the family plays a primary role in socialisation. But whether this leads to any conclusions as to the appropriateness or inappropriateness of various kinds of upbringing depends on what we want children to be socialised into, and the assumption that it should be the norm of the married heterosexual couple with children has been under challenge for some time. These issues become most clear when we discuss homosexual couples.

Homosexual couples

If we just want security of relationship, then it is argued that two homosexual men or two lesbian women could have just as committed a relationship as a heterosexual couple. But the aspect that would be lacking would be the model of a representative of either sex. Views may differ on the psychological importance of this, but it seems obvious that, as far as socialisation is concerned, it would matter only to the extent that society was modelled on the heterosexual unit as the basic unit of society. Obviously a child may suffer unhappiness if socialised into ways of life that are not only different from but disapproved of in society at large.

Single parents

One argument for allowing the single person to reproduce is that it is better, all things considered, for a child to be brought up by a single parent than by two people who are constantly quarrelling, or by a couple where physical abuse of one partner by the other is witnessed by the child. And yet society does not prevent such couples from having children.

It can be seen, again, that much of the argument depends here on views about what is the best way to bring up children, and about what children need for their proper development, and this is notoriously contentious.

Whatever the truth of the matter, however, since single-parent families are becoming more numerous anyway, as a result of marital breakdown and divorce, is there any good reason to argue against provision of children to parents who intend from the start to remain single?

Some find the deliberate planning of such a thing worse to contemplate than the unfortunate and unforeseen after-effects of divorce. But what exactly is it that is objected to here?

It might be argued that a person who takes a calculated decision to deprive the child of a parent of the opposite sex is likely to make a worse parent, if such a decision can be interpreted as showing too little regard for the welfare of the child. But cannot the same be said of deliberate decisions to seek a divorce, or deliberate decisions to remain single after widowhood, where these decisions have the result that children are denied a parent of the opposite sex? This is certainly not a view which is prevalent in our society, but what is the difference?

Perhaps the difference is that after divorce or after widowhood the children have at least access to a *history* but where a parent has deliberately decided to have a child as a single parent there may be no such history. But again, this applies to the use of e.g. AID for married couples too. This is not an issue peculiar to single parents. However, it is an important factor as we have seen.

It is clear from the above discussion that a utilitarian view will not be indifferent to the fate of the family, but must take the consequences of its possible demise into account. Thus the utilitarian will be interested in the claims that the family performs crucial social functions, and that it provides the most satisfying way to live, and the most suitable environment for the bringing up of children.

However, he will also be aware that these claims have been challenged e.g. by feminists and psychologists. Some have seen the family as the 'source of all our discontents'.[54] And a utilitarian will find it difficult to understand a view which holds that we should hang onto the family at all costs even if it is by no means a satisfying way to live. But in so far as it does enable children to flourish, he would agree with the Board for Social Responsibility. Thus R. M. Hare presents arguments in his paper to show that there need not necessarily be a conflict between Christian precepts and utilitarianism.

Feminism

The concerns of feminism in the twentieth century have embraced demands from women to control their own fertility. It might reasonably be thought, then, that feminists would favour reproductive technology as a means of providing women with more choice in this area and with the possibility of having children without marriage.

Nevertheless, the developments have caused alarm for several reasons. Some critics ask why some men have gone to such great lengths to provide remedies for infertility, and have attempted to analyse their motives. Thus the reproduction revolution could be seen as the latest expression of the long-standing attempt by men to gain control of reproduction, traditionally the woman's preserve. (The Wendy Savage case may be another example of this struggle.)

Mary O'Brien in *The politics of reproduction* argues that there have been two crucial moments in the history of reproduction. The first was the discovery of the role played by the male. This led to the restriction of women within the private sphere and to male control of women, to deal with the perennial problem of the uncertainty of paternity. Along with the banishing of women to the home went a downgrading of the private and a celebration of the importance of the public, the sphere of production as opposed to reproduction.[55]

The second momentous event was the discovery of efficient contraception, which gave women control over their own fertility. This, O'Brien thinks, changed the situation dramatically. However, it has been pointed out by writers other than O'Brien that research resulting in the contraceptive pill, carried out by men supposedly for the benefit of women, has in fact led to the expectation that women are permanently sexually available, which has not been a liberation at all.[56]

We may now add to O'Brien's two events a third, viz. reproductive technology. Again, the research that led to the birth of the first test tube baby was carried out by men. Human males have now gained the ultimate control, the power to make babies in the laboratory. In the process, uncertainty has been extended from paternity to maternity. Can a woman ever be quite sure that it is her egg, now fertilised, that is replaced in her womb?

Perhaps those who see the issues in this way and who are unhappy about it are in the minority. But from a consequentialist point of view it will be important to consider the long-term effects for relations between the sexes which may result from reproduction in the laboratory. So this perspective may not be so much an objection to a consequentialist approach as a reminder that there is another set of consequences to consider. We shall have cause to return to the issue of male–female relations in Part II.

Doing what comes naturally

To the objection that these new technologies are unnatural it is notoriously argued that if we gave up all that is unnatural we should have to give up not only all medicine but also deodorants, and surely that provides a *reductio ad absurdum* of the argument. Warnock certainly dismisses it fairly rapidly.[57] The better course may seem to be to assess each proposal in terms of benefits and costs.

However, the view that at least we need some idea of what is natural to our kind has considerable intuitive appeal. There seems to be something in the idea that there are certain modes of existence that are appropriate to our species, just as there are certain modes of existence appropriate to other species. Hence the objections to factory farming, because it denies some species these modes. We have considered the suggestion that the nuclear family might actually represent the most satisfying social unit.

Nevertheless, such claims cannot be regarded as any more than contingently probable. It may be that we could find other ways of living which we should find more satisfying than anything we have tried as yet.

What we can take of value from the natural argument is that it should be recast as a counsel. It is frequently invoked when we are introducing new and untried ideas, and in these situations, the problem is that the consequences are even more unpredictable than consequences usually are. We may find that the more a practice seems unnatural from an intuitive point of view, the greater the likelihood that it may not be worth doing, in terms of costs and benefits.

We have looked at objections to the desire-satisfaction approach, but have argued that in so far as these objections are concerned with harm to human beings, whether embryos, children, or adults, a sensitive utilitarianism must take them into account. Those who make these objections may nevertheless feel that there is more in them than can be accounted for by an assessment of consequences.

As Peter Singer has pointed out, however, the utilitarian position has in its favour that it is at least one of the things we must do, viz. examine the consequences of our actions. Whether or not more is needed, examining the effects of policies on the desires of those affected is a good place to start.[58] Anyone, according to Singer, who wants to go further and suggest that there are moral rules or definite prohibitions on certain kinds of action has the onus of proof on him of establishing what these are. Hare's paper suggests that there are none, when it comes to dealing with these novel problems.

However, even those who see the force of the argument from desire, within a utilitarian framework, acknowledge that there are severe problems in its application.

Evil desires, sour grapes and jaded palates

We have considered the existence of a very strong desire to have children, and have also looked at the possible effects of artificial reproduction on the desires of children, such as the desire to know one's genetic parentage. How are these to be weighed against one another? Even if we do not take the view that the desires of different people are incommensurable, it may be felt that the desires of as yet unactualised children will always lose out when weighed against the strong desires of adults presently existing to have those children. Also, it may be argued that those children who are produced will have *some* desire-satisfaction in their lives which will outweigh the dissatisfaction of lack of a genetic history, and which must be better than no life, and no satisfaction, at all. So if we are considering maximisation in terms of strength and immediacy of desire, the desire of adults for children will always win, as far as this calculation goes.

However there are arguments to suggest that there are other important considerations in a utilitarian calculation. Perhaps there should be some other way of *selecting* desires for satisfaction, given that it is an inevitable feature of the maximising approach that one cannot satisfy them all.

Evil desires

The fact that many people have evil desires constitutes a problem for the desire-satisfaction approach. R. M. Hare in *Moral thinking* gives the example of the ancient Roman penchant for the gladiator fight. Hare thinks this problem can be overcome by offering alternative entertainment.[59] This may be felt to be beside the point in the present context, since it would be odd to characterise the desire for a child as an evil desire (though some of the means may seem evil to some), but it brings out the point, mentioned above, that alternatives must be considered.

28

Sour grapes

A further and very serious problem is what is called the phenomenon of 'sour grapes'. Jon Elster has pointed out that many of our so-called desires are generated by what we expect and see as possible.[60] Thus if we think that it is impossible for us to achieve something we may cease to want it. Conversely, wants can be generated. This is how advertising works: it induces wants in the target audience. Many people who want videos today would not have wanted them twenty years ago. Thus it is conceivable that many of those who want children are those who have been induced to want them by stereotypes, role models, peer group pressure, etc. We considered this earlier in looking at whether the desire for a child is biological or social.

Responses to this might take one of several different forms.

One possibility is to take the view that we should try to distinguish what people 'really' want from what they say they want. The dangers in this approach are obvious. The paternalist implications are offensive to many, and even if we assess this suggestion in terms of its likely consequences, it seems clear that there will be much resentment produced by a lack of respect for people's avowed claims concerning what they want.

Reflection on this alternative may also lead us to see another possibility. Once it is admitted that many of the things that people say they want are induced by advertising or other means, certain parties may see the attractions of trying to make them want something, by means which happen to be available. This problem of 'adjustment' has been extensively discussed by Jonathan Glover in *What sort of people should there be?*[61] If people are going to get what they want, even though we have made them want it, how can a utilitarian object? After all, we should have a world of satisfied people. Is it enough to say that from our present standpoint we do not prefer such a world?

For those who find it unpalatable a third alternative is to accept people's desires at face value, and to attempt to maximise preference-satisfaction, taking people's preferences as expressed. But even this has problems. Glover has pointed out how concentration on certain desires e.g. the desire to drink espresso coffee might lead to the neglect of others, and thus to a severe loss of quality in life which we may not, on reflection, prefer.[62]

Jaded palates

It is a commonplace that to satisfy a person's desires does not necessarily produce a satisfied person, because s(he) will have new desires. 'The more you have the more you want.' Thus some may see the current demand for male pregnancies as a sure sign of the jaded palate's demand for ever more fanciful things. In the context of worldwide society, it is easy to see all claims for artificial reproduction as proof of just how pampered we are in the West.

Modifications to the desire-satisfaction approach

The part a desire plays in one's life

Jonathan Glover has suggested that a more sophisticated utilitarianism might give weight to the part played by a particular desire in one's life.[63] Thus if we came to the conclusion that the desire to beget played a more central part in a person's life than the desire to bear, we should not have to consider a man's desire to impregnate as being on a par with his desire for a male pregnancy.

How are we to judge, however, the part played by a desire in a person's life? Are we to take his word for it? The same problem seems to arise as with sour grapes.

Public policy

At the end of the Foreword, Warnock says 'The question must ultimately be what kind of society can we praise and admire? In what sort of society can we live with our conscience clear?'[64]

However, Warnock never answers this question. At a later stage we read 'We saw it as our function to concentrate on individuals rather than on the world at large.'[65]

But these wider social questions are crucial, especially when we are considering legislation. We are familiar with areas in which as a matter of public policy the law intervenes in order to place limits on desires. A well-tried example is where a majority has the desire to persecute a minority, and anti-discrimination legislation is passed.[66]

Sir David Napley's paper argues that the law's function is to protect rights and he suggests that these rights may come from two sources. They may have been established by our legal system itself, or they may fall within established principles. He makes clear exactly what rights at present covered by the law of our society may be raised in issues concerning *in vitro* fertilisation. Thus he addresses issues of legitimacy and the duty of care of the practitioner.

But he also is adamant that it is up to society to establish new rights, and that this may be necessary in this area. This is a task, he says, not for lawyers alone:

> That is not, and in my opinion never should be, a task exclusively for lawyers, whose views are as valuable, but certainly no more valuable in that connection than those of any other discipline, group or informed citizens.

It is, then, for society to decide what legal rights there should be. But how do we decide that? Interestingly enough, even Lady Warnock in her discussion of the public and the private suggested that the legal question, as opposed to the moral question, could be solved by utilitarian reasoning, by having an eye to the common good.[67] There clearly are desires, not only for individuals to have certain things, but also to live in certain sorts of society. If we give priority to the immediate desires of individuals we may end up with a society that no one wants in the long term.

And of course those who are going to be affected in the long term are the children we produce now. We need to take a policy decision about the circumstances in which children should be introduced into the world, with reference to what sort of social consequences our decisions will have for them. This is not a departure from the desire-satisfaction approach, but it is making, as a matter of policy, certain desires take priority, i.e. the desires of the children who will be produced, in assessing the appropriate response to artificial reproduction.

For it is essential to remember what is actually at issue. We are not simply talking about enabling adults to participate in an activity, like ice-skating, or to acquire certain possessions, like a new car. Speaking in terms of enabling people *to reproduce* or *to have children* sometimes disguises the reality of what is going on. We are talking about the circumstances in which new people should be born. In this context perhaps a concern for their welfare should take priority.

Of course, we cannot demand superhuman sacrifices from presently existing adults, so we need to consider alternatives for them.

The implications of this approach might be that we should give

31

heavy weighting to, for example, the desire to know one's genetic history. This would give us a reason for thinking that it is wrong to bring into the world a person who will not know who either one or both of his parents are.

Plato advocated a reproduction policy according to which parents would not know their children, nor children their parents.[68] If the idea of never being *permitted* to know who one's own children are is objectionable, then it is difficult to see why it is not also objectionable to produce children who will never be able to know who their own genetic parent(s) are.

This would not of itself rule out birth by donation, but it would imply at least that a right to know should be assigned.

However, this may not be enough. It is important not to lose sight of the biological reality. Jerome Lejeune's paper suggests the importance of confronting certain biological facts.

Biology

A child just *is* the result of the combination of the gametes of a man and a woman. Of course it can be argued that this is not all a child is. A child is not simply a genetic package but is also the product of a social environment. The Board for Social Responsibility recognises this dual element and has doubts over which is the more important of the two.

However, in the central case of reproduction there is a link between begetting and rearing. In all cases where donation is involved this is broken. It can of course be argued that the link has been broken, historically, in the case of orphans and adoption. But the reproductive revolution breaks the link as a matter of *policy*. Is this a good idea?

Arguments for the desirability of making the break have been put forward from the time of Plato's *Republic*. But Plato's scheme was designed to avoid the intense partiality of family relations which distract people from their duty to the state. The new methods of donation are designed to make artificial reproduction seem as much like the central case as possible, to enable people to satisfy their desire for children.

What will be the social effects of making the break? Apart from the challenge to the nuclear family, which we have mentioned, there is the possibility that it will encourage lack of responsibility among genetic parents for those they have begotten. Then there is the danger of unwitting incest among the future adults. Further, sociobiologists have told us about the strength of kin altruism. The less children are brought up by their genetic parents, the less there will be of this.

To some these dangers may not seem very worrying. Kin altruism is not the only form of altruism of which we are capable, even if it is the strongest.[69] The danger of incest can be minimised by regulating donation as Warnock suggests, if we really think the benefits of donation make it worthwhile.[70] But what about changes in our attitudes towards the responsibility we owe to our genetic offspring? Might not this lead to undesirable social consequences? R. M. Hare tells us that when we are considering a slippery slope, we must consider whether *in practice* it is likely. Might there be a slippery slope from donors helping out childless couples to an undermining of our sense of responsibility to our genetic offspring?

Perhaps not, if the idea of donating gametes is kept distinct from that of donating children. But to me it seems not completely *un-*likely, because these gametes are going to turn into living examples of the policy of breaking the link between begetting and rearing. Others will disagree.

But in any case there seem to be other reasons for maintaining a link between begetting and rearing, in terms of creating the best chance of a satisfactory environment for children, given the way children can feel about being deserted by their genetic parents:

> I must find my father even if it's only to discover what kind of man sells his sperm and ultimately his own flesh and blood for $25 and then walks away whistling a happy tune without any thought of the life he may have created. How is a child . . . supposed to feel about a father who sold the essence of his life so cheaply and to a total stranger?[71]

Of course this kind of feeling is not universal (perhaps she would have felt differently if no money had been involved?). Some may take the view that *these* are the attitudes that will change. The presumption that people's 'parents' are genetically related to them may disappear. This is a version of the 'sour grapes' idea, that when fewer and fewer people are genetically related to their parents, people will cease to be upset if they are not, and if they have no knowledge of their genetic roots. But what is the advantage in changing *these* attitudes, if any, rather than attitudes towards childlessness? At the moment we have some reason to believe that kin altruism is natural to our kind. We would need to be fairly confident that the benefits of weakening it would outweigh the disadvantages.

It is true that we accept the practice of adoption without worrying about its effect on the responsibility of genetic parents. But on the whole we think of adoption as second best. We can accept that adopted children can have very happy lives, but to set out deliberately to create one to be adopted seems odd, just as we accept that handicapped people can have very happy lives, yet we would not deliberately set out to handicap someone.

Alternatives

It will be objected that we are giving too little weight to the pain of childlessness, which can only be understood by those who have experienced it, and to the fact that a child born by donation will be a much loved and wanted child.

Only those who have experienced it can understand the pain of *unrequited love*, and it is no good to point out to the loved one how loved and wanted a partner (s)he would be, if that person does not consent to it. The lover simply has to look for alternatives. Some never succeed in finding a partner to replace the one they loved and lost. They may feel their lives ruined. However we do not feel obliged, as a society, to do anything about this in the way of providing partners.

The loved one's lack of consent is the crucial factor here, but the child, as Floyd and Pomerantz pointed out, cannot be consulted.[72] We have to decide, but we also have to consider what alternatives there are, as does the rejected lover.

What would be the practical implications of a policy of not allowing births by donation? Single people, and homosexual couples, would be unable to avail themselves of reproductive technology. For as we have pointed out, the biological reality demands a man's sperm and woman's egg. A homosexual couple cannot, either normally or by the technologies at present under consideration, have a child in the sense that a heterosexual couple can, as an expression of the relationship of the two of them. They can only have children by donation.

However, they are not being denied the opportunity to have children. If they are prepared to take the necessary steps ('the primitive sign of wanting is trying to get'[73]) their desire to beget can be satisfied.

R. M. Hare suggests that it may be a worse thing if people have to go to bed with each other when they do not want to, but this seems clearly to be their choice. Should we think ourselves obliged to offer IVF to a husband and wife on the ground that the sexual excitement has gone out of their marriage?

It is not a question of discriminating against homosexuals or single people on the grounds of their lifestyle or sexuality, as Sheila Maclean suggests.[74] For it is suggested that heterosexual couples should not use donation either.

Let us consider where one partner is infertile. Suppose the wife has no egg. There is the possibility of egg donation. But by denying this possibility we should not be denying her a chance to *beget*, for we *cannot* offer her that anyway. We can at most offer the chance to *rear*. Donation of sperm, or egg, or embryos, may enable individuals either to pass off, on others, children as their own or to make it appear to themselves as much like having children of their own as possible. Although the link between begetting and rearing has been broken, this may not be admitted. Thus 'prenatal' adoption may be preferred to adoption at a later stage because although only the desire to rear can be satisfied in this way, it is almost as if the desire to beget is also achieving its goal. But how satisfying is it, ultimately, to have only the appearance of the real thing?

If it actually is the case that it is only the desire to rear that is at stake, then perhaps we should concentrate on encouraging willingness to give up babies, who already exist as foetuses, for adoption, rather than aborting them.[75] We can also point out that there may be other outlets for what is involved in the desire to rear, just as the Roman spectators may have enjoyed an alternative to the gladiator fight.

The hard case for a view which opposes donation is the plight of the fertile person who is married to an infertile partner. What about the infertile woman's partner? He is capable of begetting but is being denied it.

The Board for Social Responsibility looks at this question in terms of the idea that a marriage should seek to fulfil itself through children. If this is the case then perhaps donation is a good idea. The Board is divided on this point. But those who support it are those who think the genetic component is *not* the most important in having a child. *If* it is not, why should we think that it is important

35

to help the one party in a couple who can beget, to do so, rather than, for example, trying to improve chances of rearing for both parties, since that is something they can *share*? If the genetic aspect is thought not crucial to the child, why should it be for the parent? On the other hand, if it *is* important, then it is undesirable deliberately to produce children who will be denied one genetic parent, for as we have pointed out the biological reality is that we have two genetic parents.

Changing the biological reality

Of course, it could be argued that the reality does not have to be like this. We may have evolved in a certain way, but the fact of how we have evolved does not tell us how we ought to behave. We may even now have the power to change the course of our own evolution as a species. In fact some commentators see what humans are now doing not as an interference with evolution but as part of our evolution.[76]

For example, suppose it became possible to replace sexual reproduction by an asexual method such as cloning. If each individual could have a copy of himself by cloning, this might seem to offer a way round all sorts of problems. Each individual alone would beget, and pass on all his or her genes. The desire to pass on one's genes would clearly achieve a greater degree of satisfaction than is the case with sexual reproduction. The offspring so produced would be only too clear about its genetic parentage. So where there is a desire to beget and a desire to know one's genetic parentage, cloning seems to satisfy both of these. Indeed, I have argued elsewhere that preference utilitarianism should support cloning.[77]

E. O. Wilson has argued that if we are to interfere with the course of evolution then we must be aware of the genetic costs,[78] and it is partly on these grounds that cloning has been thought to be unacceptable, even if it were practicable. Cloning might make impossible the variety we need to survive as a species.

To what extent we should be concerned with genetic costs and benefits is a question to which we shall return in Part II. For the moment we need to consider another aspect to changing our biology, and that is male pregnancies.

Male pregnancies

It would be beyond the scope of this discussion to go into the question of transsexuals, but let us consider the possibility of men who are not transsexuals having desires for male pregnancies. Is this a desire that there is a good case for satisfying?

Dr John Money of the Johns Hopkins Medical School, Baltimore, is quoted in the *Daily Mail* as saying 'Having a child is a joyful event. If it's biologically possible, why shouldn't a man be able to share in this?' Objections have tended to concentrate on the possible psychological effects on the child and the risks of death involved in the necessary procedures.[79]

Why might men want it? It might be argued that it is giving men a chance to do what the majority of women have always been able to do, viz. to bear, in other words enabling them to have an experience that they have always been denied.

Thus it could be seen as an appropriate response to feminism. Feminism has led women to encroach on what has traditionally been men's preserve, viz. the workplace. Now men are fighting back and trying to break into what has traditionally been women's preserve, viz. childbirth. It is the logical next step beyond the phenomenon of the house-husband.

But this overlooks a vital point. Feminists have made the crucial distinction between sex and gender. Thus they have argued against socially produced boundaries between women and men, which define a 'proper place' for women that in fact has no basis in sex differences.

But if we are to argue that men should be able to carry out biological functions that women can, this could be matched on the other side by, for example, an argument that women ought to have the experience of impregnation. In other words, we should be arguing for the elimination of sex differences as well as gender differences. And if we want to eliminate those, what about menstruation? Gloria Steinem's famous 'If men could menstruate'[80] imagines the consequences of this, but only on the supposition that roles were reversed. If the sexes were equal in this respect it could have enormous social consequences, which I shall not spell out here.

The vision of a society without sex differences, as in Ursula K. Le Guin's *The left hand of darkness*, has its attractions, in terms of lack of sexism. But we must ask if we need to change the biological reality in order to eliminate sexism. It would be nice to think that we do not have to change the biological reality in order to affect the way the sexes are valued and treated. But the implications of the reproductive revolution may have very serious consequences for the sexes, as we shall see further in Part II.

However, it may be that this is not the point. There are some heterosexual couples where both the man and the woman can produce gametes, but the woman cannot *bear* the child. One possibility is full surrogacy, i.e. where the sperm and the egg are joined by IVF and then a surrogate mother simply provides her womb. There is no gamete donation as in the cases considered above. One of the strongest objections to this is that the carrying mother, although not contributing her own egg, may form a bond with the child, and has in one sense a claim to be its mother, although not its genetic mother. If the man could carry the child the pregnancy could be kept in the family.

It seems the objection to this is the possibility that the child may be adversely affected, whether in other words it will miss something vital in not having been carried by a woman. A woman's *uterus* has been shown to be unnecessary by the case of a mother who had a child after a hysterectomy.[81]

Some objectors feel that the child may suffer role confusion as a result of its 'father' having been its 'mother'. But as society is becoming more used to exchange of roles between men and women it is not clear why this should be the case. If men can be house-husbands and women breadwinners, why should not men be pregnant? Children apparently can tolerate artificial incubators, so why not the incubator of a man's body? After all, as R. M. Hare points out, is not the uterus always an incubator?

There is no problem of introducing either genetic or emotional attachment from outside the couple, and so the question seems to turn on whether there is something about a woman's body that gives it elements essential to being an acceptable incubator that a man's body does not have. Similar considerations apply to the development of artificial wombs, which would enable people to avoid carrying children in their bodies altogether.

Conclusions

We have argued that it is difficult to take as a starting point for addressing these issues the claim that all adults have a right to reproduce. This right may however be assigned as the result of argument. The central argument seems to be the argument from the desire for a child, which seems to be very strongly felt, although it manifests itself in various ways and may be to a certain extent socially determined.

However, we have also argued that since not all desires can be satisfied, some form of selection must be operated. In this context it is appropriate as a matter of public policy to give priority to an assessment of what the desires of the children produced as a result of our practices will be. Whether they will be happier in a nuclear family setting is controversial, but there is evidence to suggest that at least they will want to know who their parents are, and we can gain some idea of what this is like through imagining the possibility of not knowing our own children.

The practical implications of this are that it seems wise to restrict artificial reproduction to methods that do not involve donation of genetic material. This rules out AID, egg donation, embryo donation, and partial surrogacy. It is a biological fact that single people and homosexual couples seeking artificial reproduction would require donation, so this excludes them. However, methods which do not involve this, such as IVF for heterosexual couples and the new GIFT technique, avoid these problems. So do male pregnancies, which seem to avoid the complications of full surrogacy, although we need to investigate to what extent a man's body can provide a suitable environment for a foetus in terms of consequences for the child produced.

We have argued that the provision of artificial reproduction services inevitably takes reproduction out of the private sphere and into the public. So policy decisions have to be made about its use.

At present we have a broad distinction between interference in rearing and interference in begetting. In our society, historically, there has for a long time been public intervention in the question of rearing. It is well established and much publicised in the media that we think it right to remove children from parents who abuse them. So the desire to rear is thought to carry less weight than the desire of children not to be abused, and the desire of other members of society not to tolerate this kind of behaviour. We also exercise stringent controls on adoption.

There are arguments to the effect that this kind of interference should be extended to the desire to beget. Hugh Lafollette, for example, has argued that we should introduce a system of licensing parents to have children, on the basis of predictive tests designed to pick out those who are liable to abuse their children.[82] This suggestion has gained little support. We have another principle, that one is innocent until proven guilty, and there are obvious problems in the suggestion that guilt can be predicted.

However, there is also the suggestion that we should interfere with the desire to beget on the grounds of overpopulation. Again, it has been argued that some people possess undesirable genes which they should not be allowed to pass on to children, which takes us to the subject of Part II. It has not only been argued but has informed social policy in some societies.

Some hold that all reproduction, including begetting and rearing, should be private; others hold that it should all be a matter of public concern. A third position is to draw a distinction between rearing and begetting, and to say that society, through the law, has to concern itself with child rearing and abuse. But once reproductive services are provided, begetting becomes public too.

For some the fact that begetting becomes a public concern in this way may in itself be enough to convince them that society will be happier without artificial reproduction, for they may fear that interferences with other aspects of begetting will become more likely — that, for example, there will be public concern with the *quality* of our children. This is the subject of Part II.

Notes and references

1. C. Fried, *Right and wrong* (Harvard University Press, Cambridge, Mass., 1978), pp. 110ff.
2. S. Maclean, 'The right to reproduce' in T. Campbell *et al.* (eds), *Human rights: from rhetoric to reality* (Blackwell, Oxford, 1986), pp. 99–122.
3. At a meeting of the Northern Association for Philosophy.
4. Jeremy Bentham, 'Anarchical fallacies' in J. Bowring (ed.), *The Works of Jeremy Bentham* (Edinburgh, William Tait, 1843), vol. II.
5. P. Singer and D. Wells, *The reproduction revolution* (Oxford University Press, Oxford, 1984), p. 41.
6. S. L. Floyd and D. Pomerantz, 'Is there a natural right to have children?' in J. Arthur (ed.), *Morality and moral controversies* (Prentice-Hall, Englewood Cliffs, NJ, 1981), pp. 131–8.
7. Quoted in S. Lederberg, 'State channeling of gene flow by regulation of marriage and procreation' in A. Milunsky and G. J. Annas, *Genetics and the law* (Plenum Press, New York, 1976), p. 255.
8. *The Times*, 4 Aug. 1986.
9. M. Warnock, *A question of life: the Warnock Report on human fertilisation and embryology* (Blackwell, Oxford, 1985), p. xii.
10. P. Devlin, *The enforcement of morals* (Oxford University Press, Oxford, 1965); H. L. A. Hart, *Law, liberty and morality* (Oxford University Press, Oxford, 1963).

11. Warnock, *Question of life*, p. xiii.

12. Ibid., p. xiv.

13. R. G. Edwards, 'The case for studying human embryos and their constituent tissues *in vitro*' in R. G. Edwards and J. M. Purdy, *Human conception in vitro* (Academic Press, London, 1982), pp. 371–8.

14. M. Warnock (Chairman), *Report of the Committee of Inquiry into human fertilisation and embryology* (The Warnock Report), Cmnd. 9314 (HMSO, London, 1984), paras 2.9.–2.11.

15. J. C. B. Glover, *Causing death and saving lives* (Penguin, Harmondsworth, 1977), pp. 83–5.

16. Warnock, para. 2.2.

17. 'The men who want to be Mother', *Daily Mail*, 19 May 1986.

18. Germaine Greer, *Sex and destiny: the politics of human fertility* (Secker & Warburg, London, 1984), p. 11.

19. I am using 'cloning' here in the popular sense of producing an exact copy of a presently existing adult individual.

20. Warnock, para. 8.13.

21. *NACK* 39 (1985), p. 14.

22. Warnock, para. 2.2.

23. Singer and Wells, *Reproduction revolution*, p. 67.

24. F. Lines, *Guardian*, 16 Jan. 1985.

25. Plato, *Laws*, translated by Trevor J. Saunders (Penguin, Harmondsworth, 1970), p. 183.

26. Ibid., pp. 253–6, 266–9.

27. Marilyn French, *Beyond power: women, men and morals* (Jonathan Cape, London, 1985), p. 468. She writes, 'Women may ardently desire children — especially sons — in cultures in which their own identity or well-being depends on it . . . But that wish is . . . not great enough to counter the lack of esteem for reproduction that pervades our society.'

28. Greer, *Sex and destiny*, pp. 1ff.

29. P. Ramsey, 'Shall we "Reproduce"? II: Rejoinders and future forecast', *Journal of the American Medical Association*, vol. 220, no. 11 (1972), pp. 1480–5.

30. Warnock, para. 2.2.

31. Singer and Wells, *Reproduction revolution*, p. 69.

32. Warnock, Foreword, para. 4.

33. Lines, *Guardian*, 16 Jan. 1985.

34. See Glover, *Causing death and saving lives*, pp. 69–71.

35. Warnock, Foreword, para. 4.

36. Ramsey, 'Shall we "Reproduce"?', p. 1482.

37. L. Kass, 'Babies by means of *in vitro* fertilization: unethical experiments on the unborn?', *New England Journal of Medicine*, vol. 285, no. 21 (1971) pp. 1174–9.

38. Warnock, para. 2.4.

39. Ivan Illich, *Limits to medicine — medical Nemesis: the expropriation of health* (Marion Boyars, London, 1976).

40. P. J. Toon, 'Defining "disease" — classification must be distinguished from evaluation', *Journal of Medical Ethics*, vol. 7, (1981), pp. 197–201.

41. Warnock, para. 4.28.

42. Immanuel Kant, *Groundwork of the metaphysic of morals*, translated by H. J. Paton in *The moral law* (Hutchinson, London, 1972), p. 59.

43. Warnock, Foreword, para. 5.

44. Warnock, Foreword, para. 8.

45. 'The proxy fathers: sowing the seeds of despair', *Sunday Times Magazine*, 11 April 1982, pp. 28–36. The Board for Social Responsibility refers however to evidence that suggests that AID children are happy and well balanced.

46. 'How shall a child know its parent?', *The Times*, 19 Feb. 1982.

47. R. Snowden and G. D. Mitchell, *The artificial family: a consideration of artificial insemination by donor*, (Allen & Unwin, London, 1981).

48. Warnock, para. 4.21.

49. There is of course a problem about whether and to what extent we should consider the desires of as yet unborn people. I am assuming here a version of utilitarianism that holds that we should give some weight to how future people will be affected by our actions. Sidgwick tells us that the time at which a person exists cannot affect the value of his happiness from a moral point of view — H. Sidgwick, *The methods of ethics*, 7th edn (Macmillan, London, 1907), p. 414.

50. See e.g. F. Mount, *The subversive family: an alternative history of love and marriage* (Counterpoint, London, 1983).

51. Singer and Wells, *Reproduction revolution*, p. 53.

52. *Observer*, 10 Oct. 1982.

53. Warnock, para. 2.11.

54. 'Is the family a failure?', *Observer*, 3 Oct. 1982.

55. M. O'Brien, *The politics of reproduction* (Routledge & Kegan Paul, London, 1981).

56. e.g. Greer, *Sex and destiny*, p. 144. She writes, 'For far too many women the oral contraceptive is just another version of the line of least resistance. They do not make love when they feel like it and cannot exert any influence on the frequency of sexual relations or the form they take.'

57. Warnock, paras. 2.3.–2.4.

58. P. Singer, *Practical ethics* (Cambridge University Press, Cambridge, 1979), pp. 1–13.

59. R. M. Hare, *Moral thinking: its level, method and point* (Clarendon Press, Oxford, 1981), pp. 130–46.

60. Jon Elster, 'Sour grapes — utilitarianism and the genesis of wants' in A. Sen and B. Williams (eds), *Utilitarianism and beyond* (Cambridge University Press, Cambridge, 1982), pp. 219–38.

61. J. C. B. Glover, *What sort of people should there be?* (Penguin, Harmondsworth, 1984), pp. 153–63.

62. Ibid., p. 156.

63. Ibid.

64. Warnock, Foreword, para. 8.

65. Warnock, para. 2.4.

66. Ronald Dworkin, in *Taking rights seriously* (Duckworth, London, 1977) thinks this sort of case constitutes a decisive objection to utilitarianism.

67. Warnock, *Question of life*, p. xvi.

68. Plato, *Republic*, 457d2–3.

69. cf. P. Singer, *The expanding circle* (Clarendon Press, Oxford, 1981).

70. Warnock, para. 4.26.

71. *Sunday Times Magazine*, 11 April 1982.

72. Floyd and Pomerantz, 'Is there a natural right to have children?'

73. G. E. M. Anscombe, *Intention*, 2nd edn (Blackwell, Oxford, 1972), p. 68.

74. Maclean, 'Right to reproduce', p. 111.

75. I am not suggesting that people should be coerced into giving up babies for adoption rather than aborting them. But a change of attitude so that people would feel more willing to do this could be advantageous.

76. See e.g. 'The microbes that are undermining the theory of evolution', *Guardian*, 28 Oct. 1985.

77. Ruth F. Chadwick, 'Cloning', *Philosophy*, vol. 57 (1982), pp. 201–9.

78. E. O. Wilson, *On human nature* (Harvard University Press, Cambridge, 1978), p. 198.

79. 'The men who want to be Mother', *Daily Mail*, 19 May 1986.

80. Gloria Steinem, 'If men could menstruate', in *Outrageous acts and everyday rebellions* (Fontana, London, 1984), pp. 337–40.

81. As portrayed on 'Where there's life', Yorkshire Television, 30 July 1986.

82. Hugh Lafollette, 'Licensing parents', *Philosophy and public affairs*, vol. 9, no. 2. (1980), pp. 182–97.

2

Test Tube Babies are Babies

Jerome Lejeune

The fundamentals of life

Life has a very, very long history but each individual has a very neat beginning, the moment of its conception. As it has been amply demonstrated, the whole biology of vertebrates teaches us that ancestors are united to their progeny by a continuous material link, for it is from the fertilisation of the female cell (the ovum) by the male cell (the spermatozoa) that a new member of the species will emerge.

The material link is the thread-like molecule of DNA. This ribbon, roughly one metre long, is cut into pieces (23 in man), and each segment is carefully coiled and packaged in the form of a little rod, clearly visible under the microscope, the chromosome.

As soon as the 23 maternal chromosomes encounter the 23 paternal chromosomes, the full genetic information, necessary and sufficient to spell out all the inborn qualities of the new individual, is gathered.

Just as the introduction of a minicassette into a tape recorder will allow the reproduction of a symphony so the information included in the 46 chromosomes (the minicassettes of the life music) will be deciphered by the machinery of the cytoplasm of the fertilised egg (the tape recorder), and the new being begins to express himself as soon as he has been conceived.

The fact that the baby will develop himself for another nine months inside the womb is irrelevant to this point as *in vitro* fertilisation has amply demonstrated.

44

The technicalities of fecundation

In natural conditions, the ripe ovum is expelled from the ovary by the rupture of the follicle and is recovered by the fallopian tube. Inside this tube it migrates towards the uterus and *en route* encounters the sperm which, among millions of others, will fertilise it.

At the end of the journey the fertilised egg, already dividing feverishly and organising itself in a miniscule embryo of one millimetre and a half of diameter, accommodates itself inside the uterine mucosa (nidation), around six to seven days after fertilisation has taken place. There, firmly implanting itself thanks to its chorionic villi, it will continue its growth until birth.

It is because normal fertilisation occurs in a tube, with ovum and sperm floating freely inside the liquid, that test tube babies are possible. Indeed, *in vitro* fertilisation uses a tube of glass instead of a tube of living tissue, but the process is, in other respects, identical.

Initially, artificial fertilisation outside the maternal body has been proposed to circumvent some cases of feminine sterility. It happens that sometimes the fallopian tubes are blocked, most often as a sequela of a sexually transmissible disease. In such cases, the spermatozoa cannot reach the egg nor can the egg reach the uterus. To circumvent this block, the ripe egg is taken out by laparoscopy and put into a vessel containing appropriate medium. Addition of sperm will lead to fertilisation.

The early embryo will be delicately transferred a few days later, through the cervix of the uterus, so that it can pursue its development in the womb.

All of this explains why Dr Edwards and Dr Steptoe could witness, *in vitro*, the very beginning of the exceedingly young Louise Brown whom they replaced a few days later in the womb of her mother. Thanks to the fundamentals of life already known, they were totally assured that this little berry-looking being could not be a tumour or an animal.

With the hundreds of cases already described in various countries of the world, a double evidence is now available, and for the first time, in our own species. The early human embryo is developing itself by its own virtue and it has an incredible viability.[1]

Viability outside the womb

That the early human being is fully viable outside the maternal body is not a surprise but a confirmation of general principles.

Even in ordinary conditions, with a rather simple culture medium (the fluid of the fallopian tube), the early human embryo can pursue its own destiny for days, maybe a week, and manage its own organisation. After one week implantation is a necessity but the viability of the early human being is such that even the uterine mucosa is not a prerequisite.

Up to two months implantation inside the fallopian tube is fully efficient. In these extrauterine, ectopic pregnancies the tiny human being, smaller than the thumb, is perfectly developed, the only danger being that his continuous growth would rupture dramatically the walls of the tube which cannot extend as a uterus would. Even in the extreme case of extragenital pregnancies, when the foetus anchors itself in the abdominal cavity directly on the peritoneum, the growth can be astonishingly normal for many months.

Protected by his life capsule, the zona pellucida first and, later, the amniotic bag he constructs around himself, the early human being is just as viable and autonomous as an astronaut on the moon. Refuelling with vital fluids is required from the mother ship.

A purely artificial fluid supplier has not yet been invented. But if it were ever possible complete development outside the womb would ensue. Such 'ectogenesis' would be the most proof that an embryo baby belongs to himself. If the bottle would argue that this baby is my property, no one would believe the bottle.

Time at a standstill

Careful refrigeration of living cells protecting their precarious molecular edifice is of common use for long preservation. At a very low temperature, minus 190°C in liquid nitrogen the vibration of the atoms is quite restricted. Time is suspended, so to speak.

Frozen sperm can thus be kept for years. If thawed carefully, they fully recover their fate as intrepid navigators. Banks of sperm are a common tool of industrial breeding.

The same is true for early embryos. Some mouse embryos, deep frozen and thawed, have managed after implantation to develop themselves into perfectly normal mice. No such experiment has yet been reported in our species.* Proposals are numerous although their legitimacy is at least questionable.

Twins at will

If the zona pellucida is split and the embryo cleaved in two, each mass can be inserted in a separate zona pellucida.

Identical twins have thus been produced in cattle and sheep. Some have proposed to do the same in man. Their rationale is not a 'production line' but the possibility of checking the genetic make-up of one of the twins. The scenario goes as follows. One of the twins is deep frozen until further transfer at the proper time in a recipient uterus. The other twin is allowed to grow for a while and is then examined for its chromosomal constitution, normality of growth, and its various chemical properties. The spared twin will then be transferred if the sacrificed one is declared to be all right. If it is not, the spared one will not be spared any longer. Supposedly, this procedure will give full insurance for successful childbearing even for a mother at risk due to chromosomal or genetic disease.

Simple arithmetic is not so optimistic. A success rate better than a few per cent can hardly be expected. As a mean, the egg donor should be tapped some twenty times for each successful pregnancy, an extremely heavy burden, not to speak of the 20 to 40 embryos who would not survive the whole experiment.

Wombs for hire

If properly kept in suspended life, there being no synchrony of ovarian cycles in the population, an early embryo could be transferred at any time in any recipient uterus. For example, a widow could accommodate a spared embryo fathered long ago by her departed partner. A candidate affected by an inheritable disease would welcome an egg from a healthy donor. A uterine foster mother could be hired if the true mother could not, for medical reasons, assume the pregnancy. Possibly a career woman could thus avoid the inconveniences of the pregnant state.

Surrogate pregnancy is a difficult issue. Should the foster mother be forced to give back the baby nine months later? Should

---

*Editor's note: *New Scientist*, 12 April 1984 reports such an experiment.

she be deprived of the right of voluntary termination of pregnancy if abortion is legal in her country? These questions are for lawyers. For the biologist, no matter what the avowed pretext, such practices would break the only assured link between generations. Up to now, in spite of all the uncertainties of passion, motherhood was an absolute certainty at delivery.

Sure enough the technique works in cattle. But what is good for calves and cows may not be good enough for children and mothers.

Manipulated embryos

The full viability of the early being and its striving for life allow many experiments.

The cells of two different embryos can be mixed together. They thus co-operate in the making of a compound animal called a chimera. To the best of our knowledge, no compound mouse has yet been obtained from more than three cell lines mixed together.[2] During the first cleavage of the fertilised egg there is an odd stage of three primordial cells. Maybe this three-cells stage has something to do with the individualising process.

It must be remembered that normally the zona pellucida prevents these admixtures. In a sense this bag protects our early private life. It is an open possibility that normally the human embryo hatches out of its zona pellucida only when its biological individuality is so strongly established that a chimeric accident is no longer to be feared.

But even if the mixing must be restricted to two or three cell lines, what about an 'artistic' embryo, an 'athletic' embryo, and a 'scientific' one fused together? Would not that create a kind of superman? Or, if DNA manipulation comes in, what about embryos receiving special sequences, producing exceptional endowments?

These fictional experiments do not deserve discussion. These nursery tales for grown-ups can be rejected easily. To devise a man wiser than we are we should be already wiser than we can be!

As for the proposals of manipulating embryos in order to recuperate spare parts for repairing children or adults, they are so far-fetched that no critical analysis can be made. Conceivably, grafts of stem cells could be of theoretical interest. They are already taken from voluntary donors such as a bone marrow graft for example. In any event, specialised tissues are not yet detectable in pre-implantation embryos.

A sex of choice

As Brungs stated abruptly about the advent of generalised contraception and, later, of efficient *in vitro* fertilisation, we have gone 'from sex without babies to babies without sex'. But the sex of the baby still matters.

The choice of the king is a son for the first child and often for the second. The same is true for the layman, even for the suffragettes and now the feminists. All the opinion polls give the same answer. If free choice were given a formidable excess of males would result.

Fortunately, no sieve is available to select preferentially the male sperms, carriers of the 'Y' chromosome. Pre-sexing of the embryo is also quite out of sight.

If an acceptable technique was some day available the state could not remain indifferent in the face of such a foreseeable disaster as a woman-deprived population. Not to infringe upon free choice and not to favour anyone, enormous computers would process the demands, producing optimal decisions. As demonstrated by Grouchy[3] the best equation is not too cumbersome to calculate. Toss a coin like before.

The very question

If our only goal is to help women who cannot procreate because of tubal difficulties, have we chosen the right track?

Let us return to technicalities. If, in reality, the early embryo is not an experimental material to be split, mixed, and manipulated, what is the interest of this trip of a few days in the outside world?

Dr Craft and his colleagues have already shown that the fertilised ovum can be implanted in the womb right away.[4] Could we not go even closer to physiological process? Possibly the egg could be placed in the uterus during laparoscopy with the sperm already being supplied by normal intercourse.

Why not study more closely the fluid of the fallopian tube? Would it not be the best medium for early development?

Research workers would be very wise to explore new avenues

rather than automatically following the long detour of *in vitro* fertilisation.

The future of medicine

Repeatedly, arguments have been put forward that *in vitro* fertilisation would help cure a whole array of diseases including breast cancer. But all the available evidence points toward other directions of research as shown by three recent examples.

Among the genetic scourges afflicting humanity mental retardation is the most inhumane. It deprives patients of one of the most precious parts of our patrimony, the full power of thought.

Some 10 per cent of the mentally retarded show a peculiar fragility of the 'X' chromosome. Numerous examinations have shown that this fragility can be healed if the cells are cultivated in a medium containing various chemicals; a simple vitamin or folic acid and its derivatives are especially efficient.[5] If it is added to the regimen of the patients their chromosome gap seems to disappear as well. Moreover, preliminary clinical trials show that their mental status can be ameliorated partially.[6]

The cure of the disease is not already at hand; but this is the first time that a chromosomal disease and its deleterious consequences can be attacked without resorting to science fiction devices.

Another terrible disease resulting from imperfect closure to the neural tube in embryonic life seems also to be amenable to vitamin therapy. As demonstrated by Smithells *et al.*[7] and confirmed by Laurence *et al.*,[8] vitaminotherapy, including folic acid, given in appropriate time to the mother at risk diminishes drastically the frequency of spina bifida. Here, again, no experiment on the embryo is required.

A third advance has been made on genetically transmitted anaemias.

During *in utero* life, haemoglobin is produced by an array of different genes working one after the other: the first during the embryonic stage, the second in the foetus, and the third in the whole adult life. If this last gene is mutated, an abnormal haemoglobin is made (as in thalassaemias or in sickle-cell anaemia).

It happens that the silent genes can be 'woken up' by a special chemical called azacytidine (Ley *et al.*)[9] and take advantage of this property so that the patients, rather than suffering from their abnormal adult-type haemoglobin, start producing again their normal foetal type. This type of rejuvenation could be of great significance for therapy without any manipulation of embryos or the foetus.

Aldous Huxley, Wolfgang Goethe and the newspaperman

A last question remains. Why is *in vitro* fertilisation such a fascinating issue? Although *Brave new world* is often quoted in this context, it is probably not the industrial production line of identical twins which is the key point. Aldous Huxley stressed another phenomenon. In that technological society, liberated from every taboo, the various dirty words were in current use. Nevertheless, the editors were obliged to reprint all the literature in order to remove the only incongruity which could not be pronounced, should not even be read, and was to be replaced by three points of suspension. This was the word 'mother'.

Motherhood a pure obscenity, an inversion of values, is a real danger Aldous Huxley has warned us about.

Another author, one of the greatest poets more than a century and a half ago, saw much farther. With *The damnation of Dr Faust* Goethe told of the tragic abandonment of the beloved, seduced and pregnant. In the second Faust the vision goes even deeper. After his pact with Mephistopheles Faust comes back to his old laboratory with his diabolic companion. They watch the successor of Faust producing an homunculus inside an alchemic vessel. The tiny creature escapes and floats in the air around the head of Dr Faust who, guided as he is by this strange dream baby, has definitely lost his mind but not his imagination. After impossible love with the ghost of Helen of Troy Faust finally accomplishes his goal. He builds an empire, a fully technical society, with the magic help of Mephisto. At the very end he gives his last orders — to silence the little church bell, the only one still ringing in his whole empire, and to destroy the little cabin in which Philemon and Baucis are still the paragon of human love. When the silence comes, when Mephisto returns after having burned the old lovers in their cabin, then, sorrow invades the heart of the doctor.

Poets are beyond science. They see it from far away, but they feel much more than technicians could ever grasp. In such important matters it would possibly be very profitable for scientists and legislators to read the great authors once again.

Maybe they could also rely on other writers who are much more accessible and living among us, the newspapermen. They too do not make science but see it from outside and their judgement is not taken lightly. They know that *in vitro* fertilisation fascinates their readers. One journalist understood why. Trying to convey all the significance of what was going on he coined the term 'test tube baby'. Sure enough, scientists objected — they had overlooked it — but the journalists knew better.

If any exploitation of the early human embryo is intrinsically repugnant, if everyone feels that those conducting experiments must absolutely respect these marvellously young human beings, it is for the scientific reason that a newsman discovered in an intuition of genius, test tube babies *are* babies.

Notes and references

1. J. Lejeune, 'On the beginning of human life'; Testimony before the Senate of the United States of America, Subcommittee on Separation of Powers, 13 April 1981.

2. J. de Jumeaux Grouchy, *Mosaïques, chimères et autres aléas de la féconda- tion humaine.* (Éditions MEDSI, Paris, 1980).

3. J. de Jumeaux Grouchy, *Les Nouveaux Pygmalions* (Éditions Tauthier Villars, Paris, 1973).

4. F. McLeod, I. Craft, S. Green, O. Djajanbalich, A. Vernard, H. Twigg and W. Smith, 'Birth following oocyte and sperm transfer to the uterus', *Lancet*, ii (1982), p. 773.

5. G. R. Sutherland, 'Heritable fragile sites on human chromosomes (1) Factors affecting expression in lymphocyte culture', *American Journal of Human Genetics*, vol. 31 (1979), pp. 125–35.

6. J. Lejeune, 'Le Metabolisme des monocarbons et le syndrome de l'X fragile', *Bull. Acad. Nat. Méd.* (1981), p. 165.

7. R. W. Smithells, S. Sheppard, C. J. Schorah, M. J. Selle, N. C. Nevin, R. Harris, A. P. Read and D. W. Fieldink, 'Possible prevention of neural tube defects by periconceptional vitamin supplementation', *Lancet*, i (1980), pp. 339–40.

8. K. M. Laurence, N. James, N. H. Miller, G. B. Tennant and H. Campbell, 'Double-blind randomized controlled trial on folate treatment before conception to prevent recurrence of neural tube defects', *Brit. Med. J.*, vol. 282 (1981), pp. 1509–11.

9. T. J. Ley, J. Desimone, N. P. Anagnov, G. H. Keller, B. Humphries, P. M. Turner, N. S. Young, P. Heller and A. W. Nienhuis, '5-azacytidine selectively increases 8-globulin synthesis in a patient with B + thalassemia', *New England Journal of Medicine*, vol. 307 (1982), pp. 1475–9.

3
Marriage and the Family

Board for Social Responsibility

We must begin by setting out what we are and are not doing in this chapter. We are *not* giving a comprehensive account of the Christian understanding of marriage and of the family. We have a much more limited aim: to consider the Christian understanding of the nature and place of procreation in marriage and the possible meaning of this for the moral acceptability of the use of the new technologies aimed at overcoming the problems of childlessness. There are many other aspects of marriage which have been and doubtless will continue to be considered by the Church which are beyond our terms of reference.

It has been traditional in the Western Church to refer to three 'ends' or 'goods' of marriage. St Augustine, with whom the tradition seems to have originated, used to speak of the three as 'offspring, fidelity and sacrament'. *The book of common prayer*, which has made the tradition familiar to Anglicans, lists them as 'the procreation of children . . . a remedy against sin . . . [and] the mutual society, help and comfort which the one ought to have of the other'. The union of two people in the completeness of marriage involving sexual, social, emotional and relational aspects, is seen as promoting the three central goods in human life: namely, the transmission of life in the human community, a disciplined structure of living in which the individual may grow to moral maturity, and a strong and enduring relationship between them. In short we may speak of the 'procreational', 'moral' and 'relational' goods of marriage.[1]

The tradition clearly takes a sympathetic view of attempts to relieve childlessness. Such attempts do, after all, assist marriage to fulfil one of its natural ends. There is a long-standing Christian concern that marriage, wherever possible, should lead to parenthood. Childless couples, though their marriage is perfectly valid and is often rich in other virtues, have been disappointed of a good which is proper to them. First, we can see how this tradition will foster a concern that the relief of childlessness should be undertaken only within marriage and as a service to marriage. 'The procreation of children' is meant in a broad sense: 'bringing them up in accordance with God's will, to his praise and glory' (ASB). The transmission of life in the human community ought not to be viewed in a narrowly biological sense. There is a social aspect, adequate or inadequate, passed on with life itself, and this is of crucial importance to the human growth and development of the child. No service is done to procreation if it is taken out of the context of family life in which the Creator wills young human beings to flourish. There is a third implication concerned with this tradition. This is concerned with *holding together* the procreational and relational 'goods' of marriage. This requires some careful examination because it has a profound implication for different ways of resolving childlessness.

We need first to explain what we mean by the use of the interchangeable terms 'end' and 'good'. Here we are using the idea of an 'end' as a purpose rather than a conclusion. The term 'purpose' may be used in two ways or may mean two different things. It may be a goal which we decide upon. This is something we fix upon by the exercise of our own freedom of choice. Second a purpose may be seen as a 'good' which is part of the order of things and something we may discern by our reason and accept as part of the given nature of the way things are meant to be. Christians think of these 'goods' as the Creator's purposes for what He has made. It is in this second sense of purpose that we talk of the three 'goods' of marriage. When we enter into marriage we enter an institution given to us by God with these three good ends as part of its nature and meaning. Christians have traditionally believed that these goods should be held together.

When it is said that the procreational and relational goods of marriage should be held together, that may sometimes mean that no act of intercourse within marriage should preclude the possibility of procreation, and that no act of procreation should be performed independently of the physical intercourse of the partners. Though this view has certainly been held, and probably still is held, within some Christian Churches, it is not generally accepted by Anglicans. Anglicans usually hold that love may be

expressed through sexual intercourse even when the use of contraceptives prevents the possibility of procreation, and that it is
quite proper to plan both the number and timing of children
within a marriage. Thus, although one may be unwilling to say
that it is entirely up to the partners whether to have children or not
within a proper Christian marriage, one may certainly say that
they may plan the number of children by the use of contraceptive
techniques. This means, of course, that an element of conscious
responsible planning necessarily enters into the matter of procreation. It therefore becomes impossible to say that the procreational and relational goods of marriage must be held together
on every occasion, or even on most occasions, since, after the
wanted number of children has been achieved, one may use contraceptives on all subsequent occasions.

Is there any way, then, in which we may say that these goods are
held together? One may say that the procreation of children is not
just an optional matter of choice for Christians. It is a proper good
of marriage, intended by the ordinance of the Creator himself. It is
one of the proper purposes of marriage. So, except for very good
reasons, every Christian marriage should seek to fulfil itself by the
procreation of children. And such procreation should not take
place outside the marriage bond. Similarly, mutual companionship, help and comfort, for better or worse, which the couple
promise at their marriage to give and accept one from the other, is
not some sort of optional extra to marriage. It, too, is a proper
good, ordained by God. Its proper locus is marriage. In saying
that these goods should be held together, one may be saying that
the intention to have children — even when it is responsibly
planned, and therefore very much brought under human control
— and the intention to maintain life-long fidelity, for better or
worse, should positively interrelate. Each should strengthen and
support the other. One may go further and say that even where
family planning is used, the partners are to accept a child as a gift
within their marriage (including children conceived 'by mistake').
And the mutual relationship should be widened to include love for
children of the marriage. In other words, the important points are:
that procreation should not occur entirely outside the loving
relationship; and that the loving relationship should issue in the
good of children, unless there are strong reasons to the contrary
(like genetic defect of a grave kind).

If it is agreed that the procreational and relational goods of
marriage are sufficiently held together in the marriage as a whole,

and that therefore contraception is not destructive of this, a further question must be asked concerning the permissibility of the use of technical procedures which do not involve any relational good (i.e. there is no physical intercourse, even less a strong and enduring union, between the genetic procreators). The issue will scarcely arise for those who have already accepted that intercourse may properly take place without leaving open the possibility of pro-creation. For they are already committed to the permissibility — and indeed the desirability — of sundering procreative and relationship acts in particular cases; and not just in a few particular cases, as we have seen, but in the vast majority. There is clearly no barrier for them to using an artificial technique of procreation, as long as such techniques are not used entirely outside the context of a loving relationship. Now in such cases, the technique is offered as an aid to the restoration of a good proper to the marriage, which through some handicap has been impeded. So it is calculated to strengthen some relational good, and the bond between the various goods which go together to make a proper Christian marriage. It seems, then, that the use of assisted fertilisation by a couple who cannot, or who are advised on medical grounds, not to have children is acceptable, since it may be said to hold together the procreational and relational goods within the marriage as a whole. (Further questions are raised by IVF which are concerned with the question of experimentation on embryos.)

It is thought that we may be attempting to achieve a mastery over human nature itself, possibly involving a reduction of it to the status of an object to be made and manipulated, in encouraging a technological way of thinking about procreation. The natural processes embody and express much larger patterns and relation-ships on which our whole experience of the world and each other depends. It is clearly possible for a mature and thoughtful couple to use a technical procedure in procreation without coming to think any differently about each other and about their children than they would otherwise have done. What is feared is the impact on our culture of a technological way of thinking about sexual intercourse and procreation. Those who feel this strongly will be reluctant to embark on such a procedure. They feel that sexual intercourse forms the centre of a network of instinctive family relationships which is complex and deep-rooted, and that nothing should be countenanced which threatens this complex network. They will feel that it is in sexual intercourse that we serve all three goods of marriage at once. It is there supremely that we not only engender children, we not only delight in our partners, we not only experience the disciplined direction of our instincts, but we do all these things together and at once. Here we grasp the multivalent structure of marriage immediately, and our attitudes to partner and children are formed accordingly.

A Christian couple may decide that, if it is permissible to plan responsibly the number and timing of children by the use of contraceptives, then we are already seeking and achieving a greater mastery over the processes of reproduction without reducing anything to the status of an object. They will know that it is not true that in each act of sexual intercourse they engender children as well as delighting each other. And so they will not hesitate in situations where they are not otherwise able to have children of their own to engender children by artificial means, within the context of their own loving relationship. They will certainly wish to guard against any undermining of commitment to the goods of marriage which, they believe, have been willed by God himself. Yet the responsible use of IVF to remove the disability of childlessness within marriage will not threaten to undermine the interweaving of procreational and relational goods in general within marriage. In fact, in specific marriages, it may offer an enrichment of the marriage relationship which both partners gladly accept.

The situation is significantly different when we come to consider AID. For here there is a question of introducing genetic material from outside the union altogether. What, then, is left of the procreative end of the marriage? Simply the parental nurture of the child, beginning with the pregnancy and birth. One parent makes a genetic, and usually gestatory contribution to the child, but the other does not. So here procreation is separated from relationship completely, at the genetic level, even though the connection between the two is preserved at the social level. In attempting to assess how serious an objection this poses to the practice of gamete donation, we may first consider two contrasting ways in which the practice is sometimes described, even though these descriptions are inaccurate and misleading.

Gamete donation is sometimes described as a form of adoption, sometimes as a form of adultery. It resembles adoption in that the parents accept the child voluntarily, although they are not its genetic parents. But of course in AID the parents take on a much more consciously responsible role. They do not decide to nurture a child who already exists. They decide to bring a child into existence who will be genetically only partly their own. Moreover, in

English law, the adopted child has the right to know its genetic parents, under certain carefully phased conditions. In AID this is not, at present, the case since the donor is almost always anonymous. Is it right or wrong to decide to bring a child into being who will not know its genetic origin? It is true that many children will not know their true genetic origins in any case, but that does not resolve the moral problem of whether it is desirable. There will be a difference of view here between those who think that the genetic origins of a child are fundamentally important and those who believe that what is more important is the loving nurture of the child in a stable marital relationship. The authors of this report, though they are all sensitive to this problem, cannot agree on the moral status of the free decision to bring a child into being with the assistance of donated gametes. There are two differing points of view held among us. One is that if donation takes place within a stable marital relationship, it still has the status of a good, though not, of course, one which should become a norm; the other believes that the perils to marriage, as understood by Christians, are so grave that the extension of gamete donation should be strongly discouraged, and AID dislodged from the established position it now holds among the techniques of aided fertilisation.

AID may be compared to adultery in so far as the presence of the child is founded on a genetic union that is extrinsic to the family. Of course, there is no offence against the married partner, there is no breaking of the relationship of physical fidelity and there is no real relationship with a person outside the marriage. It is certainly quite unlike the physical act of adultery therefore.[2] But the parents have still decided to bring into being a child who is not genetically their own, and which does involve the procreative (even if anonymous) activity of another human being. Is it right or wrong to decide to bring into being a child who is not genetically the offspring of partners concerned in the decision? We are conscious that Anglican tradition hitherto has opposed the donation of genetic material.[3] However, it is right to reflect again on the adequacy of statements about this in the light of further knowledge and experience. Bearing in mind that we are only considering cases where marriages are deprived of the good of children, what we are asking is whether this defect can be remedied by the use of genetic material from outside the marriage. We differ on this, depending on whether we see the genetic as the most basic manifestation of the personal and find the alienation of genetic parenthood from marriage a development which undermines the

Christian understanding; or whether we judge that, although everyone is fundamentally influenced and limited by his or her genetic endowment, nevertheless the overriding factor is the social context which can assure proper love, respect and care. To this extent the question of genetic origin is not of fundamental moral importance, when compared with the question of how the child will be loved and cared for.[4]

Another concern is that the extension of human decision-making about procreation beyond the genetic confines of the married couple introduces an element of dominion over nature which appears unjustifiable to some and possibly even threatening to human values. But to others it is little more than an extension of that responsible control over procreation which contraception already has introduced. Though there may be dangers in human control of procreation, they are unlikely to be realised in these very limited and carefully controlled situations, so that what is required is the existence of safeguards to the procedure, rather than its prohibition. Even if some feel personally uneasy about the use of AID, they may not wish to prohibit its use in law, by those who conscientiously feel that it may strengthen the marriage bond and remedy the lack of a great and natural good of marriage. (We refer specifically to the safeguards set out in the Board of Social Responsibility's Response to Warnock in paragraph 5:4.)[5]

Family affections stretch back to and embrace pregnancy, during which the bond between mother and child — and, at one remove, between father and child — is emotionally secured. This fact leads us, despite our disagreements on ovum and semen donation, to a common mind on the practice of surrogate motherhood. This term refers, strictly speaking, not to a technological practice (for the technology is the same as that of gamete donation with IVF) but to a contractual or quasi-contractual agreement. The parents, having made their genetic contribution to the fertilisation of the embryo *in vitro*, will assume the duties and privileges of parenthood only after the child has been carried in the womb by another woman. Our view is that the case for ovum donation itself — i.e. for the separation of the female genetic contribution from the mother's gestatory role — stands or falls with the claim that the genetic relation is in itself not of decisive importance compared with the gestatory one. But in the case of surrogate motherhood, it is precisely the gestatory role which is being minimised. To those of us who accept a comparative devaluation of genetic factors on their own, it seems clear that the mother who bears the child has a true parental, and already to some degree a social, role. Thus any subsequent transfer of parental responsibility must be viewed as a form of adoption arrangement. And, while there is nothing wrong with adoption, there is something undesirable about creating children specifically for the purpose of adoption, and even more so, about creating them for a financial consideration. To those of us who in any case regard the genetic contribution as of overriding importance, the separation of the two female contributions to the biological origins of a child appears as in itself unacceptable.

We are in accord, then, that in surrogate motherhood the Christian institution of the family is fundamentally endangered, and thus that it cannot be morally acceptable as a practice for Christians. We therefore associate ourselves with the strong recommendations on this subject made by the Warnock Report. On other types of gamete donation we are not of one mind, but we would all wish to ensure that such practices were properly controlled and recorded. We would like to see the anonymity of donors qualified in some way, with a considerably improved system of record-keeping and perhaps with certain information about the donor (though not his or her identity) made available to the child when it comes of age.

Finally, we would wish to reiterate that our fundamental concern in these matters is for the preservation of the good of Christian marriage, as instituted by God himself, and for the welfare of children, who are to be brought up in the fear and love of the Lord. It will need much observation and discussion before we can come to a clear mind about whether these practices threaten marriage or the true welfare of children, or conversely if they are a blessing to marriage. But it is above all important to recognise the new situation in which we stand, with possibilities now open to us which have never before existed. In this situation, our traditions of moral thought need to be extended and rethought. It may well be that previous ways of thinking will not be sustainable on reflection. On the other hand, we should not give up too lightly positions which have been important to generations of Christians. Our working party has contained committed Christians who span a wide range of Anglican moral traditions. We have sought to provide a guide to thinking on these issues which has been rather painfully forged in our own discussions. We believe that it is important to continue to wrestle with these questions, and we reiterate the welcome of the Board for Social

Responsibility for a national licensing authority

to regulate research and to control infertility services. We believe that this body should also continue the debate on the moral aspects of technologies concerned with human embryology and fertilisation, including new technologies which may become possible. We welcome the proposals made in the Report concerning its composition, and in particular the recommendation for strong lay involvement. In addition to the categories of membership mentioned in the Report, *we recommend that there should also be adequate representation from the social work and legal professions, and from members of the churches skilled in moral theology* (para. 1.1.).

Notes and references

1. These emphases are different in other traditions. For example, the Book of Common Order of the Church of Scotland lists the goods as 'the lifelong companionship, help and comfort which husband and wife ought to have to each other . . . the continuance of the holy ordinance of family life . . . and the welfare of human society'.

2. A case in the Scottish courts has declared that AID is not legally adultery. In Edinburgh in 1958 a Mrs Maclennan successfully defended herself against a charge of adultery by claiming that a child born to her was the result of AID. The judge ruled that physical intercourse was necessary for adultery to take place.

3. See the findings of the Archbishop of Canterbury's Commission in 1948, and the 1959 Memorandum of Evidence submitted on behalf of the Church of England on AID to the interdepartmental committee chaired by Lord Feversham.

4. The limited number of studies of children born following AID that have been carried out suggest that they are generally more balanced and healthy than the average. After all, they have been desired and sought in a way which is not true of some children born of normal intercourse.

5. 'We agree to all the proposals of the Committee of Inquiry concerning the availability of AID on a properly organised basis, and would wish to strengthen some of their recommendations:

(a) We think it inappropriate that donors should sell their semen for gain. Semen is not a commodity to be bought and sold: it is the God-given means of making possible the gift of new human life. If people are willing to donate their blood free of charge, suitable people will be willing to donate their semen. Indeed, donor is something of a misnomer in the present situation. The Report acknowledges some of the objections to payment, and proposes a gradual move to a system whereby semen donors are paid only their expenses. *We recommend that, if AID is made available on a properly organised basis, expenses only should be paid from the outset.*

(b) We welcome the Report's recommendations that a child by AID should be recognised as a legitimate child of the marriage, and that at the age of 18 the child should have access to information concerning the donor's ethnic origin and genetic health and possibly some other basic information. We also welcome the view that 'the sense that a secret exists may undermine the whole network of family relationships' (p. 21). Those of us who favour the use of donation in defined circumstances believe that openness from the start is a help both to couples and to children in establishing good family relationships. We hope that this will be impressed on parents by the statutory counselling service which we have already recommended (3.3). We are aware that there is a lack of knowledge about children born from AID and there is an urgent need for research into the needs of such children. *We recommend that this research be undertaken.* Only on the basis of such work can the exact needs of these children be established and the nature of the provision that is required to meet these needs.'

4

IVF and the Law

Sir David Napley

The task which has been assigned to me is to endeavour to indicate what interests the law should protect in relation to *in vitro* fertilisation. I shall doubtless be forgiven if I seek to limit my remarks to the task thus assigned. *In vitro* fertilisation is concerned, as the title conveys, only with fertilisation in glass, i.e. test tube embryo produced *extra utero*.

This can again be limited to four specific procedures. The first involves the extraction of ova from a woman; the impregnation of it, within a laboratory, by her lawful husband's sperm and the return of the fertile ovum to the mother. The second involves the same process save that the sperm is provided by a male other than the husband. The third is that in which the ovum is extracted from a third party, and, after fertilisation by a husband's sperm, is introduced into his wife's body in order that she may in due course produce a child. The last possibility is where third parties provide both the ovum and the sperm which is then inserted into the potential mother's womb.

It is, I believe, important, against that background, clearly to understand what is, and should be, the effective purpose and objective of a legal system. The law, in any country, exists to protect rights. Those rights may be such as have been established by Parliament or any other law giving authority or which have been recognised under the Common Law, or other law enforcing process, as having so long existed and been so long recognised that they cannot now be gainsaid. The only other rights which the law recognises are those which, whilst not hitherto delineated in detail, fall within established principles, such as those relating to the achievement of justice, or sound and fair government and the like.

Indeed, it is, itself, a principle of the English Common Law that the courts, even when defining a right which has not previously been declared or known to exist, never create new law but merely declare what the law, albeit unremarked, has always been. Thus, in relation to *in vitro* fertilisation or any other subject, the law can only protect rights which can be said already to exist in the manner which I have described. The law does not ever, on analysis, truly protect interests, therefore, but only the right to them.

· When one turns to consider which rights, currently having no place in society, ought to exist, we move into the field of legislators, who should, but certainly do not always, endeavour, in a democracy, to reflect the will and wishes of the people. That is not, and in my opinion should never be, a task exclusively for lawyers, whose views are as valuable, but certainly no more valuable in that connection than those of any other discipline, group or informed citizens.

The problems arising from the development of *in vitro* fertilisation it will be seen are today almost entirely concerned with what rights ought to exist, and far less with any which already exist, so I and other lawyers have little specialised wisdom to contribute.

Fertilisation involving the ova and sperm of a lawfully married couple gives rise to no greater number of problems than ordinarily arise from procreation in its normal form.

God, or nature, depending on your theological viewpoint, wisely made procreation attractive by linking it with a highly pleasurable and sensual process. It has done something similar, albeit less compelling, for the ingestion of food, unless you had the misfortune to marry a poor cook.

If, from necessity or otherwise, you choose to sustain yourself on a balanced diet with a series of pills, you may just as effectively maintain your health, although denying yourself the pleasures normally associated with eating. The same conditions apply to procreation.

If, at the private or public level, both are freely consenting parties, no one enjoys a right to prevent you denying yourself pleasure. If one or neither consents to the process wholly different considerations obtain. If a fertilised ovum, whether her own or someone else's, or sperm, whether her husband's or another's, is artificially introduced into a woman's body without her consent this constitutes an assault and battery at law and is criminally and civilly actionable as such. Moreover, if it involves wounding or serious injury it may constitute an even greater crime or cause of action.

What, however, if a man's sperm were to be used without his consent to fertilise an ovum *in vitro* or *per uterus*?

What remedy, if any, does the law afford him? There is, presumably, no property in sperm, and unless the person acquired possession of it under a contract, there could be no cause of action for unauthorised use. This may well be an aspect where the law should, at least, be amended to give the producer of the sperm the sole right to determine any future use to which it may be put.

It has never been a ground for dissolution of marriage that both parties agreed to refrain from sexual intercourse and even a decree of nullity, on the ground of non-consummation, is, I suspect — since I do not believe it has ever been precisely decided — based upon the denial of the opportunity to procreate rather than the denial of sexual pleasure. Certainly the church has opposed contraception on the basis that intercourse which is undertaken for pleasure, with the opportunity of procreation being eliminated, is sinful and forbidden.

Similarly, there is hardly justification for denying a right to inherit on intestacy or to succeed to an hereditary title because although conceived within the framework of both the ovum and semen of lawfully married parents, it occurred without the physical sexual act which is otherwise essential. No one has, so far, suggested that the marital state of the parents as generated by love or affection has a positive effect on the genes, chromosomes or the structure of ova or sperm. If and when they do we may have to reconsider the legal situation.

There remain in this connection, however, other questions for resolution. The first flows from medical or scientific negligence and is a matter which, being common over the whole field of practice, may better be dealt with later. Another derives from the necessary practice of extracting a number of ova from the female so that some remain surplus to the operation. The only legal right to which that gives rise is the right of individuals to use their bodies as they choose unless in the process they violate rights already vouchsafed to others. Thus, in this country there is no law denying a woman freedom to be a prostitute, only laws which regulate the extent to which she may inconvenience others in the process. What happens in the future to her ova or his sperm is, in law, entirely a matter for her or him.

It is therefore inappropriate to ask what the law should protect.

The only answer is that the law should protect the existing rights of the public or the individual. The relevant question is what rights should the public or the citizen possess which they lack, and that is a question not of law but of morality, theology and what either legislators, the public or both decide is the best for the future of society. Although I may have opinions on that, my views are not exclusively entertained as a lawyer and it must fall to others, mistakenly or correctly, to express their opinions and reach their decisions without regard to my views. When the legislative has decided these virtually insoluble problems the law can and will protect the rights which result.

The remaining processes of *in vitro* fertilisation give rise to some difficult legal problems. Should a child whose origins were not derived at all, or only in part, from its lawfully wedded 'parents' — in inverted commas — inherit property or title on the same basis as a child conceived in the normal way in lawful wedlock?

The law in this country now draws no distinction in terms of inheritance on intestacy, in respect of illegitimate children since the Family Law Reform Act 1960, or adopted children, since the Adoption Act of 1976. However, inheritance on intestacy has drawn a distinction, over many years, whereby, in certain circumstances, it defers issue of half blood to issue of the whole blood. The expression 'half' and 'whole' 'blood' may, in the modern context, seem curious to the biologist but would seem equally to have application where the ovum or sperm was provided by the wife or husband of the lawful marriage in which the child is nurtured. More difficult is the question where neither ovum nor sperm was so derived. Yet if 'illegitimate' means conceived out of lawful wedlock then the conception manifestly takes place in the laboratory and not, one would believe 'in wedlock'. Once again whether that should be regarded as immaterial is a moral theological-social-medical or general public opinion problem and the law can provide neither protection nor decisions until it is resolved by others.

Fears have been expressed of a high level of risk of abnormal children being produced by laboratory fertilisation. If this occurs what remedies does, and should, the law provide?

If, as appears to be the practice, doctors or scientists, or both, contract with a husband and wife, or one of them, to fertilise ova by one or other of the processes, the doctor or scientist clearly undertakes a duty of care to the party or parties with whom he has contracted. If, as a result of negligence, damage is sustained by them in relation to the abnormality of the child or in any other manner which flows directly from the negligence, they clearly have a remedy available at law.

What however of a claim by the foetus itself or the child once it is born? Since Parliament passed the Congenital Disabilities (Civil Liability) Act of 1976 a person responsible for an occurrence affecting the parent of a child which causes that child to be born disabled is liable in damages to the child if he would have been liable in tort to the parent. This however, does not extend to create liability for a pre-conceptual event if, before conception, the parents knew of, and accepted, the particular risk. It might be thought that the liability in relation to *in vitro* fertilisation should be further extended.

If doctors or scientists are to assume the mantle of God and create life (and, with stronger reason, if unlike God they exact a fee for their services) they should, perhaps, assume responsibility for ensuring that no disability will flow which the exercise of reasonable care and skill could have avoided. They cannot, where things go awry, expect to be protected, as is God himself in such cases, by the insuperable difficulty of serving him with a writ. Moreover, that liability ought in this special field to extend totally to what occurs pre-fertilisation as well as post. Additionally, although the courts have held (*McKay* v. *Essex Area Health Authority* (1982) 2 W.L.R.890) that this recent statute does not give a child a right of action for 'wrongful entry into life', legislation should afford such relief where, especially in return for a fee, doctors or scientists facilitate just that very occurrence.

In the field of Family Law there are problems to be solved. The father is the natural guardian of a legitimate child but the mother occupies that position in relation to an illegitimate child.

What is the status of a child fertilised *in vitro* from the ovum of a woman rather than the one who delivered it, and from sperm provided by someone other than her husband? Since the process is not unlawful it is difficult to call the child 'illegitimate'.

As I have observed, it seems equally fallacious to regard it as having been conceived in wedlock. The increasing tendency in Family Law is to look to the membership of the family rather than to the genetic origins. If the process has occurred with the consent of the husband and wife and the fertilised ovum has been implanted within the wife, commonsense and the support of the concept of the family would seem to dictate that the law should invest the child with all the rights and privileges of lawful offspring. That, at least, is the view of the Law Commission. Those who would afford rights to the man who provided sperm, or the woman the ovum, are surely stretching credulity and rationality to unacceptable limits.

The law can hardly be defined as to the situation of an unmarried woman who gives birth as a result of artificial means, until society has first determined whether it wishes to proscribe, discourage, permit or encourage such conduct. This necessarily involves such questions as the mores of society itself, the future of such a child, the likely effect on its mental, psychological and social well-being and innumerable other considerations which the law alone cannot determine.

A further question which legislation alone can resolve is what restrictions (if any) should be imposed on further developments in this field. Many will consider that the world will be made no happier by the total development in a laboratory of human beings, but lawyers alone cannot hold back the coming of the Brave New World, which seems likely in this atomic age to replace the Frightened Old One.

Parliament again can alone decide whether it is to be the crime of abortion to terminate the life of an embryo fertilised *in vitro* and not yet implanted in the woman. The philosophy underlying the abortion law is not the protection of the mother but the prevention of the destruction of life. Theoretically, there can be no distinction between foetuses *in* or *extra utero*.

The scientist who fertilised the ovum would seem to have no greater right to terminate the growth than a mother or father who normally conceives, and their right is a most limited one.

Yet common sense again seems to indicate that a proper distinction can and should be made, but it cannot be made until society has made up its mind about the wider issues. In like cases are the questions whether surrogate motherhood — for commercial or other reasons — should be protected, restricted or proscribed although few can doubt that once an *in vitro* fertilised child has been born and established in a family no other parental claims ought properly to be recognised.

It will, by now, have become evident that the question is far less what are the interests which the law should protect so much as what are the rights which should be enacted so that the law may protect them.

Without any claim that, as a lawyer, I am the best person to prescribe such rights the following are some of such as may in summary seem worthy of consideration.

(1) That a child born as a result of *in vitro* fertilisation with the ova or sperm of the lawful spouse into whose family it is delivered shall be declared to enjoy all those rights and privileges which would accrue to a natural child of that family.

(2) That it should be a serious criminal offence to implant a fertilised ovum in a woman without her consent or by fear, fraud or duress.

(3) That no one should be permitted to fertilise without the consent, freely given, of both the provider of the ova and the sperm and that it shall be a serious criminal offence to act in breach of this dictate.

(4) That it shall be forbidden — and a serious criminal offence — to provide ova or sperm or, in the case of a woman, her body to facilitate fertilisation, in return for money or other valuable consideration, other than the reimbursement of reasonable and necessary expense incurred.

(5) That the right to engage in the process of *in vitro* fertilisation be limited to properly qualified scientists or doctors of a specific degree of experience and training.

(6) That the parties on whose behalf the operator undertakes to achieve such fertilisation and the child when born shall enjoy an unrestricted right to recover all such damages as flow from the fact that the child when delivered is abnormal or disabled unless the operator can establish, on a balance of probabilities (the onus of proof being on him) that such abnormality or disability was not occasioned by any failure on his part to exercise reasonable skill and care.

The onus should be upon the operator because the difficulties of proof if resting on the proposed parent or parents would be so insuperable as to nullify the rights so created.

(7) That in the interests of society it shall be a criminal offence, carrying condign punishment, to fertilise and develop to maturity (should it appear to be possible) an embryo outside of the body of a woman.

(8) That fertilisation outside of marriage should, again, in the interests of society, be forbidden under pain of substantial penalties.

(9) That all right of inheritance and title should, for all purposes, derive from having been established for a specific period of time as a member of the family and without regard to the method of procreation employed.

(10) That it shall not amount to abortion or otherwise be

unlawful to terminate the growth of an embryo, at any time before it has been implanted in the host-parent, with the consent of the spouses for whom the process was begun.

Whether a list is comprehensive of all that is required and whether or not you agree with the thoughts expressed, it is sufficient to add additional fuel to the fire of the debate.

5

In Vitro Fertilisation and the Warnock Report[1]

R. M. Hare

The philosophical contribution to discussions like that about *in vitro* fertilisation (IVF) and related problems should be to help sort out the good from the bad arguments that are used. One has only to read the many pronouncements on this subject in the newspapers to realise that what is needed above all is a sound and generally accepted method of argumentation, armed with which those who start with different views can discuss them with one another in the light of the medical facts and possibilities, and in the end, we hope, reach agreement. It is adherence to a sound method of argument that will bring this about; in default of it the discussions can go on inconclusively for ever. Since nearly all the questions involved are moral or ethical, it is the moral philosopher, the specialist in ethics in the narrow philosophical sense, who should, if he is good at his trade, be able to help here. The help consists not in handing down conclusions but in enabling others to reach them by sound arguments.

Moral philosophers differ from one another about their subject; there is not just one accepted theory in moral philosophy any more than there is in any other field. I have to try to say what I think is correct, and at the same time not to stray too far from what would be generally accepted by those in my profession who understand the issues.

Let me now draw attention to some features of the problems raised by IVF, and to one in particular. This is that they are quite new problems. Until a few years ago hardly anybody envisaged as possibilities what are now actualities; and this makes us take very seriously the problems presented by what are now only informed speculations about what might become possible. The danger is that these possibilities might be realised before our moral thinking was ready for them. This is indeed what has already happened. Since the problems are new, we ought to be cautious in applying old precepts to them. These precepts got generally accepted when things were very different from what they are now or may become.

I will start, just to illustrate this point, with a fairly simple example, before I go on to some more contentious ones. There is a generally accepted condemnation of adultery. I say 'generally accepted'; it is not, of course, universally accepted, because many *avant-garde* people have come to think that there is nothing wrong in it. But let us leave them on one side and just consider the position of those who still think that adultery is wrong. There is an argument that could be, and is, addressed to such people, which goes like this. Adultery is wrong, as we all agree. But adultery is *defined* as the union of the sperm of a man with an ovum of a woman when they are not married, and one of them is married to someone else. That is what adultery essentially consists in; the means adopted is not essential. Hitherto only one means has been available, namely sexual intercourse. But if new means, such as artificial insemination by donor (AID) or some forms of IVF become available, they do not alter the essence of the act; it is still adultery and therefore, as we all agree, wrong.

There is more than one fault in this argument. One obvious objection that is likely to be made is that that is not what adultery is: a necessary element in it is sexual intercourse. If that objection is accepted, the argument that AID or IVF by donor is adultery and therefore wrong collapses; for they do not involve sexual intercourse. So the union of sperm and ovum is not a *sufficient* condition of adultery. But it is not a *necessary* condition either. A union would be adulterous even if contraception were successfully used, so that there was no union of sperm and ovum.

But a less obvious, yet more fundamental, fault in the argument is that it gives no reason for the ban on adultery. If a reason were given for it (as I think it can be given), it might turn out that the reason had its basis in circumstances and conditions which used to hold universally but now no longer hold. And if the conditions no longer hold in a certain kind of novel case, it *may* be, or may not be (we cannot say without further discussion) that the reason does not hold any longer either. We can only settle this question by a closer examination of the reasons why adultery has hitherto been thought wrong.

If I were asked what *my* reasons are for thinking it wrong, I would say something like this. I am firmly convinced, unlike some of the *avant-garde* people I mentioned just now, that marriage is a valuable institution for the happiness of the partners and their children. Though it sometimes goes wrong, I am convinced that in our human situation no other system for the procreation, nurture and upbringing of children is likely to do nearly so well. I emphasise that on this point my views are completely orthodox and conventional. So they are on the next point, that adultery is one of the greatest dangers to the stability of marriages, and therefore to the happiness of the partners and their success in bringing up their children happily and well. I will not here inquire why it is a danger. The sociobiologists will perhaps tell us that there are genetic reasons why in humans the marriage bond is normally a firm one. That is why anything that endangers it is commonly the subject of severe condemnation. Adultery, because it diminishes trust and for other reasons, does so endanger the marriage bond. Let us, anyway, suppose that this condemnation is justified by the evil consequences that adultery normally has in weakening the marriage bond, not only for the particular partners, but in general.

So far I am in agreement with the general condemnation of adultery. Why then do I not accept the argument I have just outlined, which starts from this condemnation as a premiss, adds nothing but a definition of adultery, plus some medical facts about AID and IVF which are not in dispute, and ends with the conclusion that AID and IVF by donor are adultery and therefore wrong?

The answer is that, even if we conceded that AID came within the definition of adultery, in such a case the reasons which made adultery wrong in the ordinary case would not apply. The ban on adultery was a general rule, a very valuable and important one, whose firm acceptance and hence almost universal observance did the best for people's well-being in nearly all ordinary cases. It could be, even before the invention of AID and IVF, that cases arose in which adultery would be the best way of furthering the purposes of marriage. There have been societies in which, if a man was infertile, his brother or some near relative was expected to secure children for his wife, children who would be members of his family, by begetting them for him. The service done for Abraham by Hagar provides a somewhat similar example.[2] If this was done with the knowledge and approval of all, I can see nothing wrong in such a custom, as a good means of securing the ends which marriage serves — among them those so beautifully set out in the Prayer Book: 'for the procreation of children, to be brought up in the fear and nurture of the Lord, and to the praise of his holy Name'.

What AID and IVF can do is to make such solutions readily available to an infertile couple, without any sexual intercourse (such as might be thought objectionable) by any but them. If there is agreement between the partners, this may be a way of cementing the marriage by providing children for them. Therefore the reasons, which I said I accepted, for the general condemnation of adultery do not apply in this special case. We can go on accepting the general condemnation, but say nevertheless that it applies only *in general*, not universally, and that AID and IVF cases, with mutual consent, can be treated as an exception to it, because in those cases the reasons for the general condemnation are absent, and there are good reasons for making the exception.

It may now be suggested that it is dangerous to allow exceptions to the general ban on adultery, because that will weaken it, and so lead to a widespread relaxation of moral standards in relation to marriage as an institution. Note that this is a practical argument, and needs to be answered on practical grounds. It belongs to a type of argument commonly known as the slippery slope, or the thin end of the wedge. It is certainly true that there are many good rules of conduct which we treat as inviolable or at least very sacred, and are alarmed by anything which tends to weaken their hold on people. Examples are rules of honesty and rules forbidding violence. If a few people start cheating other people, and are not condemned and brought to book, there is likely to be a general collapse of honesty and nobody will be able to trust anybody else. If a few people who feel like it, and are strong enough, violently assault their neighbours in pursuit of some advantage for themselves, and get away with it, others will soon copy them and anarchy will be the result.

I must stress again that this is a practical argument. For it to succeed, it is necessary not merely to show that *in theory*, if the fraud or the violence were condoned in a particular case, widespread fraud or violence in more unusual kinds of case could ensue. It is necessary to show that, as the world is, it is *likely* to ensue. There are cases in which the principles forbidding fraud and violence are with general consent relaxed. The most notable of these is where the police use force or deceit, in ways permitted by the law, in order to catch criminals; nobody says they must not because, if *that* fraud or violence is allowed, fraud and violence will spread in society. What should be permitted to the police in the way of fraud or deceit can be, and is, disputed; but nobody thinks that they should be forbidden altogether.

It is perfectly practicable to have exceptions to well-established principles, provided that the cases in which the exceptions are allowed are well demarcated and easily recognisable. I want to suggest that the use of AID and IVF to secure children for a childless couple might be such a case. We might allow it, on clearly specified conditions, without in any way weakening the prohibition on adultery in the general case. There are two ways in which we might describe this exception to the general rule, and it is important to be clear that, though I shall give a reason for preferring one of these ways, we could choose either without making any substantial difference to the argument. We could say either that, though the definition of 'adultery' remains the same, and AID is adultery, it is allowable when it is without intercourse and with the consent of the other spouse in order to secure a family; or we could say that in such a case the union of sperm and ovum is not adultery. Similarly, in the case of the police, we could say that the use of force within the law is not violence (it is *vis* but not *violentia*), or that it is justified violence. And we could say either that the use of deceit to catch criminals is not fraud, or that it is justified fraud. It does not matter.

I mention this point because in the discussion of such questions there is a danger that a lot of time will be wasted in disputes about whether such and such an act is really adultery, or really murder, just as it has been on disputes on whether the embryo or the foetus is really a human being. What matters is whether a certain act, about whose nature we are quite well enough informed, is right or wrong, not what we are to call it. Philosophers are, among other things, concerned with the use of words: they should use such skill as they may have acquired to *prevent* our wasting time on disputes about words, not to bog us down more deeply in them.

The reason why people get bogged down in such verbal disputes is that they have some general principle, the reasons for which they cannot give, and want to hang on to it at all costs ('Always keep a hold of Nurse for fear of finding something worse'[3]). From this point of view, if you can persuade yourself that AID is not really adultery, you can go on holding the principle that all adultery is wrong, and still allow AID when it seems right to you to do so. But if your original principle was a reasoned one, you can, without any detriment to it, say that, although AID is, strictly speaking, adultery, the *reasons* for condemning adultery in general do not apply in this particular case. On the whole I do not recommend this second way of speaking, because 'adultery' has become fairly well accepted in our language as a word implying condemnation, as Aristotle noticed;[4] therefore if we called the donor an adulterer it would be difficult to avoid giving the impression that we were somehow condemning him, which, if we approve of what he has done, we do not mean to do. But all the same the verbal question is not important enough to detain us for long.

I want now to sum up the lessons to be drawn from this relatively easy case for our understanding of the right method of argument about such questions. We may then be able to handle more difficult questions with greater hope of settling them. The method I advocate goes in the following steps:

(1) Where a new practice looks as if it might be a breach of an old principle, but otherwise seems to have a lot to recommend it, first ask what were the reasons for the old principle. They will be found to consist in certain good consequences that come from the general acceptance of the principle, and certain evils that would ensue on its general abandonment.

(2) Ask, next, whether in the new case these same reasons still hold. If they do not, there is the beginning of an argument for relaxing the principle in such a case.

(3) Ask, then, whether such a relaxation is likely to lead *in practice* to a general weakening of the hold that the good old principle has on people's attitudes and behaviour. This may depend on whether the type of case in which we want to relax it can be easily recognised and demarcated from the cases in which we want to keep the prohibition.

(4) If the answers to the last two questions are favourable — if, that is to say, the reasons for the old principle do not hold in the new case and the making of an exception will not in practice put us on to any slippery slope — then we can make an exception of the new case, if the situation resulting from the acceptance of the principle with the exception written into it is better than that resulting if we retain the old principle unmodified.

I have been speaking throughout in terms of moral principles; but the same will hold if we are talking about laws. In the AID case and the analogous IVF case the question was one of adultery, and adultery has in Britain not usually been treated as a crime, though it often has in America and in Muslim countries, and has had consequences in civil law even in Britain, for example in divorce cases. That was why I dealt with it as a moral, not a legal matter. Nevertheless, even the cases I have so far considered can become, and have become, matters for proposed legislation; and the same applies *a fortiori* to some other uses of IVF.

However, I think that I need not say anything more about laws over and above what I have said about morality. This is not because law and morality are the same thing — far from it — but because in this area the reasons for, and the arguments for, changes in the law are closely analogous to the reasons for changes in our moral attitudes. Not all sins should be crimes; but if you are wondering whether to make some new practice into a crime, the questions you should ask yourself are rather like those which should guide you if you are thinking of treating it as a sin. There are differences into which I shall not have time to go. But in general the question we have to ask ourselves in both cases is, 'If we changed our law or our moral attitudes, would the new state of affairs be preferable to the old, from the point of view of all those affected?'

The approach which I have been recommending has been, broadly speaking, a utilitarian one.[5] I have adopted it because it is the only approach which seems to me to yield clear answers to questions such as we are faced with, and because the answers which it gives are, I am sure, those that would commend themselves to anybody who had a firm understanding of the questions he was asking and of the facts. Utilitarianism is, however, not a universally accepted doctrine among philosophers, and it sometimes arouses the hostility of theologians. I wish to try to mollify anti-utilitarian philosophers and theologians by showing that all the same things can be said in their different languages.

Some such anti-utilitarians may wish to say that the good old principles, like that forbidding adultery, should be sacrosanct; we ought not to question them. Intuitionist philosophers will say that we know them by intuition. Theologians may say that we know them by revelation either in scripture, or to the church or its members by the Holy Spirit. I will take each of these contentions in turn and show that, properly understood, I do not need to maintain anything substantially different from them.

As regards moral intuitions, a place can easily be found for them within a utilitarian system such as has been guiding me. It is certainly a good thing that we have them, and do not normally question them — that we are very firmly convinced that some kinds of act, like adultery in the ordinary case, are wrong. The same applies to the *feelings* on which Lady Warnock lays so much stress in her thinking. It *is* important that we have feelings of outrage and shock, as Sir Stuart Hampshire calls them,[6] or intolerance, indignation and disgust, to use Lord Devlin's expression.[7]

But there are two different reasons, one good and one bad, why these reactions might be thought important, which need to be distinguished, as I think the Warnock Report does not. The first is that it is one of the facts of life that certain things shock people. Legislators have to take account of such facts. If a lot of people are shocked by what happens as a result of some piece of legislation, then there may be adverse consequences for society. Devlin was quite right about that; and it is obvious that *if*, as was not the case, the consequence of liberalising the laws about homosexuality had been a breakdown of public morals and of respect for the law, Devlin would have had a good *utilitarian* reason for deploring such a reform (and he does generally speaking argue on such utilitarian grounds).

The other, bad, reason for thinking such reactions important is the thought that they might have some value as premisses in our moral reasoning. I mean that the *opinions* might have value, not the fact that they are current. On this view, we are to argue, not from the premiss that a lot of people *think* homosexuality wrong but from the premiss that it *is* wrong. But the fact that those feelings exist does very little to establish *that*. The most that it establishes is, if people have been thinking about the matter for generations, and are well informed and not muddle-headed, that they may have by trial and error come to some sensible opinions. I have at least *some* confidence in the wisdom of the ages. But it may be doubted whether the old prejudices about homosexuality have this standing. And when we are arguing about wholly new problems like surrogacy by IVF, this is manifestly not the case. To judge by the rapidity with which people changed even their inveterate opinions about homosexuality that Devlin took so much for granted, it would be most unsafe to rely on the Clapham commuter as a moral guide. It would only be safe if she had been thinking deeply about the matter and was well informed. It is very important to have moral reactions, but still more important to have the ones we ought to have.

To come back to my first example: the difference between an

intuitionist and a utilitarian does not consist in the first holding that adultery is wrong and the second holding that it is not wrong. It consists, rather, in the first being unable or unwilling to give any cogent reasons for his conviction that it is wrong, whereas the second is able to show why it is a good thing that we should have this conviction and seek to cultivate it in others, including the children that we bring up. The reason is that the state of affairs in which adultery is condemned is better than that in which it is condoned or encouraged. Naturally I think that a philosopher who is able to give cogent reasons for his convictions is superior to one who is unable or unwilling to do so; but, although they disagree about method, they need not disagree on any point of substance.

In facing the theologians I am more anxious, because I am not at all well versed in their arcane disciplines. But all the same I think that my utilitarian approach can be amply justified in religious terms. I take as my starting point the Golden Rule, common to all the great religions, that we should do to others as we wish that they should do to us (that is to say, in identical circumstances, 'identical' being taken to include the mental states and dispositions of those affected, so that someone is not required to go round whipping people because he himself likes being whipped). The Golden Rule has the same effect as the injunction to love one's neighbour as oneself. Both of these are really consequences of the other great commandment to love God. If we believe that God himself loves his creatures, and therefore wills their good impartially, we shall think that to show our love of God is to strive to do his will by loving our neighbours as ourselves, seeking their good as we do our own, and treating their good as of equal weight to ours, and therefore to each other's; and that is the utilitarian doctrine.

However, in our human condition of ignorance, lack of time for thought, proneness to self-deception, and sheer confusion of mind, it is impossible for us to determine reliably, on particular occasions, what is God's will — that is, what would do the best for all our neighbours, showing equal love to them all as God loves them all equally. So, inescapably, we have to be ruled by more particular commandments, since we cannot apply the great commandment directly. These above all are the content of what is called 'conscience'.

One of the wisest of the English philosopher-divines, Bishop Butler, expressed extremely well the relation, about which it is easy to be confused, between the great commandment of love and the particular commandments, for example the commandment not to commit adultery. He said:

> From hence it is manifest that the common virtues and the common vices of mankind, may be traced up to benevolence, or the want of it. And this entitles the precept, *Thou shalt love thy neighbour as thyself*, to the preeminence given it, and is a justification of the Apostle's assertion, that all other commandments are comprehended in it; whatever cautions and restrictions there are, which might require to be considered, if we were able to state particularly and at length, what is virtue and right behaviour in mankind. For instance [he adds in a footnote], as we are not competent judges, what is upon the whole for the good of the world; there may be other immediate ends appointed to us to pursue, besides that one of doing good, or producing happiness. Though the good of creation be the only end of the Author of it, yet he may have laid us under particular obligations, which we may discern and feel ourselves under, quite distinct from a perception, that the observance or violation of them is for the happiness or misery of our fellow-creatures. (Sermon XII)

Note the appeal to St Paul;[8] he supplies the authority both of scripture and of the church that I said earlier might be relied on by those who do not wish, or do not feel able, to think rationally for themselves. There is more about this in Butler's *Dissertation on virtue*,[9] which should be read by all who are interested in the relation between benevolence and the particular virtues and obligations. These latter, he thought, are revealed to us by our consciences.

But this leaves a gap in our thinking which theologians do not do enough to fill. The *content* of what our consciences tell us (like the intuitions which are the same thing under another name) is not constant from one person to another. It is partly the result of their different upbringings. We bring up our children to (as it is sometimes put) 'know the difference between right and wrong'; but what they think right and what they think wrong may be influenced by what in particular we teach them. So, especially when we are thinking about novel questions like the present one, we sometimes do not know what to teach them, or what to think ourselves, and conscience gives us no clear guidance. We inquire (as best we may, and I am not pretending that it is easy) 'What would a being, who could do this kind of thinking better than we can, be likely to want us to think? What ought our moral convictions to be?' And we shall have to decide this question, as best we may, in the light of what we think God's purposes are. We can take it that he wishes good to all impartially, so we can only ask ourselves, what will secure this. And so we find ourselves asking, as the utilitarians also bid us ask, what moral attitudes, convictions, and the like will best serve the purpose of doing good to all those affected, treating them all as equally objects of our love. So I do not think that there ought to be any conflict on this matter between the best theology and the best philosophy.

I have illustrated the questions before us, and the method we should adopt in settling them, by taking one of the easiest. If we could grasp the right method in relation to this relatively easy question, we might then be able to go on to handle more difficult ones. I mean questions like 'Is it wrong, and is it murder, to propagate human embryos by IVF and then use them for purposes of experimentation or transplants?' and 'Ought we to allow women to bear other women's children for payment?' I have opinions on all these questions, arrived at by the same method as I have been describing; but I shall not discuss them directly now, because I want to go on first to apply what I have said about methods of argument about these matters to an assessment of the Warnock Report. In the course of it I shall refer to two of these questions, but shall not have time to defend my opinions about them. Actually my opinions about surrogacy are not in general very different from those of Peter Singer and Deane Wells in their excellent book on this subject;[10] so I can refer to it for a defence of them. The views of the two dissenters in the Warnock Report are quite similar.

I suppose that there are two obvious alternative lines that philosophers in a committee like Mary Warnock's can take. One is to try to get the committee to clarify the issues before it, bringing into the light of day the arguments on both sides, assessing them, and then making clear the reasons which have led the committee to its own conclusions. I realise that this sounds rather like what some idealistic enlightenment figure from the eighteenth century might have said:

As steals the morn upon the night
And melts the shades away;
So truth does fancy's charm dissolve,
And rising reason puts to flight

The fumes that did the mind involve,
Restoring intellectual day.[11]

Certainly Mary Warnock seems to have abjured any such
ambitions. She was content with a second best alternative, which
was perhaps all she could manage. This was to find some conclusions which the members of the committee, or as large a
majority of them as possible, would sign, and not bother too much
about finding defensible reasons for them. Since the members
were fairly typical in their moral attitudes or prejudices, it might
be hoped that conclusions to which they could agree would also be
acceptable to the public, and that government policy would duly
conform. It is in such matters easier to get people to agree about
conclusions than about the reasons for them.

On most of the matters the committee discussed, this policy on
the part of Mary Warnock proved, from the purely political point
of view, absolutely sound. The Committee had a very good press;
initially its recommendations seemed to be welcomed by the
government and the public, and legislation to implement them was
in the offing. There was, however, more dispute about two of the
main recommendations: to permit some experimentation on
embryos up to two weeks, and to forbid professional or administra-
tive assistance, whether commercial or non-commercial, for
surrogate motherhood, and make contracts for surrogacy unen-
forceable in the courts, but not to ban it altogether.

I should like to observe in passing that the consequences of forbidding professional assistance for surrogacy are illustrated by an
actual case recently reported in the papers: two people who did not
in the least want to go to bed with one another had to do so in order
to achieve a surrogate pregnancy, and this would have been quite
legal under the Warnock recommendations, since nobody
arranged it but the people involved. And, as is argued and illustrated at length in Singer and Wells's book, nearly all the troubles
with surrogacy as it is beginning to be practised in America would
be overcome by having a *public* (state or licensed) agency to
manage it, somewhat as is done for adoption in many countries.

The public, in a field that it had not yet thought much about,
was in general inclined to accept the judgement of the good and
great on the committee; but these two recommendations about
surrogacy and embryo research seem to have aroused more
unease. The result was the recent Enoch Powell Bill, which was a
great deal less liberal than the Warnock recommendations, and,

supported by Cardinal Basil Hume in *The Times*,[12] very nearly got through Parliament, where it commanded a large majority and was only stopped by a procedural manœuvre. The subsequent trend of legislation has been illiberal and, as I think, irrational.

What this showed was that Members of Parliament, like the public, had not thought nearly enough about these difficult and novel problems, which had only emerged in very recent years, and which nobody had studied enough to clarify the issues even to those working in the field, let alone make the public clear about them. It may seem too severe to say so, but *this was what the Warnock Committee should have been doing*. What was missing in the public discussions was any understanding of the arguments on one side or the other of these questions. In default of reasons, people fell back on their prejudices.

Why was the Warnock Committee unable to do much about this? There are of course the difficulties I have mentioned already; to get members of the committee to study, and even agree in the clear statement of, the arguments on both sides, a lot of hard *philosophical* work would have been required; and, according to Michael Lockwood, Mary Warnock has said that this proved unpalatable to the committee.[13] But I do not think it should have been impossible to do better. I have myself served on a number of much less exalted committees on similar subjects, most of them set up by the Church of England[14] (which anybody who did not know the very enlightened colleagues I had on them might think an unpromising sponsor). Actually I believe we did some good. One of these committees produced its report just before the debates on the 1967 Abortion Act which introduced the reasonably liberal regime that now governs abortion in Britain. I think that the passage of the Act was helped by our Report, and especially by the active part which the chairman of our committee, the late Bishop Ian Ramsey, played in the House of Lords in getting many of the bishops to support the Bill. Another committee might be said to have influenced the church's views on the treatment of terminal illness.[15]

But I do not need to speak of my own experience, which is not so grand as Mary Warnock's. We have, in the Report of Bernard Williams's Committee on Obscenity and Film Censorship,[16] a very good example of what can be done by a philosopher who is determined to get such a committee to think rationally. It would be hard to better the chapter in that Report called 'Harms?', in which the costs and benefits of restrictive and permissive legislation about various kinds of pornography are very well discussed. It must be admitted that, perhaps *because* it was so enlightened, the Williams Committee did not get nearly such a good press as the Warnock Committee, and its recommendations were not taken up. But it may have in the end a more lasting influence, because, although with changes in the pornography scene the Williams Report may go out of date in some respects, the reasons are all there and will eventually be absorbed.

If we want a *successful* example of a more ambitious approach, we have only to look at the Wolfenden Committee's Report on homosexuality.[17] This was published many years ago when prejudice against homosexuals was pretty strong; but the Report succeeded in actually influencing public opinion, or at least crystallising it in a way which made a very sweeping reform of the law a possibility, so that now we have what is on the whole a fairly satisfactory law on adult homosexuality. This was what Devlin was protesting against; and perhaps Lady Warnock's committee, like Lord Devlin, will prove to have been voicing, on the questions of surrogacy and embryo research, opinions typical of her own generation, which may in the twenty-first century come to seem archaic.

Why were the Williams and the Wolfenden Committees able to give reasons for their recommendations, whereas on these controversial questions the Warnock Committee gives hardly any, and certainly few that will stand up to any sustained attack? The explanation is a philosophical one. Both Wolfenden and Williams (surprisingly in the light of his well-known anti-utilitarian views) argued in a consistently utilitarian way. The chapter in the Williams Report called 'Harms?' is, I said, a careful and extremely well worked out cost-benefit analysis of various legislative proposals. The procedure of the Wolfenden Committee was similar. It seriously tried to discover what would actually happen if various policies were followed, and what impact this would have on the interests of the various parties affected, including members of the public. These were the reasons it gave.

There is hardly anything of this sort in the Warnock Report. We are simply told how the Committee feels, and how other people feel, about surrogacy and embryo research. The reason why the committee was not able to help in giving very solid reasons for these 'feelings' is that she is not a utilitarian, and all the reasons that in the end will hold water are utilitarian ones. Her committee ought to have been asking what would happen if, for example, embryo research was forbidden or allowed, or surrogacy was

permitted under various restrictions. I mean, what would happen
to those affected: the people who would be helped by the research;
even the animals who would otherwise have to be sacrificed in
alternative research; the potential people into whom the embryos
would turn if it were possible to implant them; and so on. But
instead, they were content on the whole with recording their own
reactions to various proposals, which, no doubt, they hoped the
public would share. This was intuitionism in action. The committee was moved by its feelings of 'outrage and shock', 'indignation and disgust'.

It may be that most of what shocked it shocks me too, so I can
accept many of its recommendations. But when it turned out that
some things shocked Enoch Powell and a large number of Members of Parliament which did not shock the committee, what could
she say? She wrote a piece in *The Times* condemning the 'absolutism' of Powell and his supporters.[18] But the absolutism was not
the real source of the trouble. It is, admittedly, important to distinguish between two sorts of intuitionists: those whose intuitions
support absolute prohibitions couched in very simple and general
terms, and those whose intuitions allow them to qualify their moral
view almost *ad libitum* according to how they feel about particular
matters. But because neither of these parties is able to give any
reasons for its views beyond deploying more intuitions, the
'absolutists' are in a much stronger position than the other kind of
intuitionists like Mary Warnock, at any rate when it comes to the
propaganda battle. They can claim to be defending simple sound
principles against erosion; and indeed the principles often are
sound ones, though only in general. However, as I think Mary
Warnock sees, the whole issue in this area is whether, in view of
the differences between the ordinary circumstances for which the
sound principles were framed and the extraordinary and novel
possibilities with which these new techniques confront us, we
ought to make exceptions to the principles. I have already illustrated this with regard to IVF and adultery.

So, if intuitions are all that are allowed to count, the absolutists
are likely to have a big advantage. What are needed to counter the
rhetoric of the absolutists are much more detailed arguments
giving the pros and cons of exceptions to the general principles.
But the Warnock Committee is extremely cursory in doing this,
and the 'reasons' it gives are often no more than expressions of
moral conviction without any support. The support would take the
form of predictions of actual harms that would come from the

introduction of the practices condemned. If the committee just does not *like* the idea of the practices, what force has that?

On surrogacy, for example, some arguments are given against allowing it. They are larded with such expressions as 'inconsistent with human dignity', 'treat [the uterus] as an incubator' (is it not always an incubator, if you choose to use that word?) and 'the child will have been bought for money'. Then some arguments are given on the other side, using such expressions as 'a deliberate and thoughtful act of generosity'; but also usefully countering some of the excesses of the first lot of arguments. And the main benefit from allowing surrogacy is mentioned: it 'offers to some couples their only chance of having a child genetically related to one or both of them. In particular, it may be the only way that the husband of an infertile couple can have a child.'[19]

When the Committee comes to its own recommendations, one would expect it to have assessed these various arguments and said which of them were good ones, as I am sure the last one was. But what it actually says is

> The moral and social objections to surrogacy have weighed heavily with us. In the first place, we are all agreed that surrogacy for convenience alone, that is where a woman is physically capable of bearing a child but does not wish to undergo pregnancy, is totally ethically unacceptable [well, maybe, but no reason is given]. Even in compelling medical circumstances the danger of exploitation [a highly emotive word used in contentions which have already been rebutted in the preceding argument] of one human being by another appears to the majority of us to outweigh the potential benefit. That people should treat others as a means to their own ends, however desirable the consequences, must always be liable to moral objection. Such treatment of one person by another becomes positively exploitative when financial interests are involved.

So, in the end, the Committee comes down on one side without giving any but the most sketchy reasons. If the risk to the surrogate mother is a reason, does that make it equally exploitative to employ steeplejacks if one pays them well? What differentiates surrogate motherhood from other forms of personal service? Maybe something does, but we have not been told what it is. Why am I not 'treating others as means to my own ends' whenever I employ anybody? In case Lady Warnock should invoke Kant here I must point out that (as is well known), what Kant said was not that one may not use people as means — he never said *that* was always liable to moral objection, and he would have been absurd if he had — but that one may not employ them *merely* as means, but always also as ends.[20] If I employ a steeplejack I am using him as a means, but also as an end, because I enable him, by paying him a high wage, to realise his own end of having good money to live on. It may be that surrogacy is different, but we are not told in what respect.

I have not room even to sketch my difficulties with what the Warnock Committee says about embryo research as briefly as I have in the matter of surrogacy. The Committee makes one extremely important and valuable philosophical move, thus locating itself in the centre of the moral problem instead of remaining on the periphery as so many do. This is where it says:

> Although the questions of when life or personhood begin appear to be questions of fact susceptible of straightforward answers, we hold that the answers to such questions in fact are complex amalgams of factual and moral judgments. Instead of trying to answer these questions directly we have therefore gone straight to the question of *how it is right to treat the human embryo*. We have considered what status ought to be accorded to the human embryo, and the answer we give must necessarily be in terms of ethical or moral principles.

To have got *that* into a public document is, I think, Mary Warnock's chief claim to a place in the philosophical Hall of Fame.

However, when the Committee comes to discuss *what* moral principles, we are again left without any but the thinnest of reasons for adopting one principle rather than another; and this is, as before, because the Committee, relying on its intuitions, is in no position to give any. For example, it says 'Everyone agrees that it is completely unacceptable to make use of a child or an adult as the subject of a research procedure which may cause harm or death', and it is then asked whether this applies to embryos. But I should be surprised if many people do in fact agree with this sweeping condemnation of research on humans. Taking a blood sample by finger pricking *may*, in very exceptional circumstances, cause harm or even death, but, given informed or even, in the case of children, proxy consent, it is a common and accepted experimental procedure in research (e.g. on atmospheric lead pollution, to determine how much lead is getting into people's blood). One of the unofficial working parties I have sat on has produced a report on the question of when experimentation on human children is allowable and when not, with reasons given.[21]

The Committee mentions the argument that the embryo is not a potential person because, unless it implants, it 'has no potential for development'. But if we are in a position to implant it, as normally we are if we really want to and are prepared to find a recipient mother, the embryo *does* have the potential and *is* a potential person. The whole question of what moral bearing this potentiality has, has been extensively discussed in the literature following the denial by Michael Tooley that potentiality has moral relevance;[22] the treatment by the Committee is most inadequate. My own view is that potentiality *is* morally relevant, but that, as is argued by Singer and Wells, it is equally relevant in the case of as yet uncombined gametes; and that, since there is an almost infinite potentiality for producing human beings *ad libitum*, we cannot have a duty to produce one from any particular pair of gametes or embryo. But I will not go into this question at length.

In general, the Committee gives no cogent reasons for its recommendation that experimentation on embryos should be allowed up to two weeks but not thereafter. What was needed was a thorough examination of the consequences for all those affected. These include, besides parents and possible parents, the potential person into whom an embryo might if implanted turn, *and* the potential people that might be brought into being if *other* embryos were implanted, or *other* pairs of gametes turned into embryos. And we must remember that research on embryos might make this possible in cases where it is not now possible. In short, they should have enquired into the good and harm that would come from allowing or forbidding such research. But this, perhaps, they were not ready to do, because Mary Warnock had abjured utilitarianism and consequentialism (which is the only rational and sensible way of handling these questions), and persuaded them to keep off such topics so far as they decently could. So we have in the Report *some* enquiry into consequences, but it is not nearly far reaching enough; and instead we have plenty of appeals to intuitions, i.e. prejudices. And that is one reason why Enoch Powell had such a field day, and the public is still floundering, and we are getting rather bad legislation on this question.

Mary Warnock is a far better politician than I am, and therefore a better judge of the art of the possible; and it may be that she *could not*, in the circumstances of her committee, do any better. If so, she is not to be blamed. But Wolfenden did, and got the law changed, and Williams did, though he failed at that time to get the law changed. And in the humbler church committees in which I served, we did reach liberal conclusions, in rather unfavourable circumstances, on abortion and euthanasia, and, as I have said, influenced legislation and the thinking of some church leaders. Admittedly, I have not since then been asked to serve on any other church committees. But I just have a lingering doubt as to whether, if Mary Warnock's *philosophical* views had been more as I think they should be, she might not have produced a more effective report.

Notes and references

1. M. Warnock (Chairman), *Report of the Committee of Inquiry into Human Fertilisation and Embryology* (The Warnock Report), Cmnd. 9314 (HMSO, London, 1984). Reprinted as *A question of life*, with new introduction and conclusion (Blackwell, Oxford, 1985).

The first part of this paper was read to a colloquium sponsored by Mr M. I. Mogul in Grosvenor House, London in 1983. The full paper was read to another colloquium in Oxford in 1985 organised by the Hastings Center and by the Oxford External Studies Department. I derived a great deal of help in my thinking from both colloquia.

2. For the story of Abraham and Hagar, see Genesis 16.

3. Hilaire Belloc, 'Jim, who ran away from his Nurse and was eaten by a Lion', in *Cautionary tales for children* (London, 1908).

4. Aristotle, *Nicomachean ethics*, 1107a 11.

5. For a defence of my own version of utilitarianism, see my *Moral thinking: its levels, method and point* (Oxford University Press, Oxford, 1981).

6. S. Hampshire, 'Morality and pessimism', 1972, reprinted in his *Public and private morality* (Cambridge University Press, Cambridge, 1978).

7. P. Devlin (Lord Devlin), *The enforcement of morals* (Oxford University Press, Oxford, 1965), pp. viii, 17. The original lecture is discussed by H. L. A. Hart, *Law, liberty and morality* (Oxford University Press, Oxford, 1963).

8. St Paul, Epistle to the Romans 13:9.

9. *Sermon XII* (1726) and *Dissertation on virtue* (1736) are reprinted in D. D. Raphael (ed.), *British Moralists 1650–1800* (2 vols, Oxford University Press, Oxford, 1969), vol. 1.

10. P. Singer and D. Wells, *The reproduction revolution: new ways of making babies* (Oxford University Press, Oxford, 1984).

11. The last aria from Handel's *L'Allegro, il Penseroso ed il Moderato* (libretto by Charles Jennens).

12. B. Hume, *The Times*, 6 June 1985, p. 12.

13. M. Lockwood, *Moral dilemmas in medicine* (Oxford University Press, Oxford, 1985), pp. 1f.

14. The Reports of the Church of England committees on medical questions are obtainable, if in print, from the Church Information Office, Church House, Dean's Yard, Westminster, London SW1.

15. The conclusions of the church committee's report *On dying well* were partly accepted by Archbishop Coggan; see D. Coggan, *On dying and dying well*, Edwin Stevens Lecture 1976 (Royal Society of Medicine, London, 1977).

16. *Report of Committee on Obscenity and Film Censorship* (The Williams Report), Cmnd. 7772 (HMSO, London, 1980).

17. *Report of Committee on Homosexual Offences* (The Wolfenden Report), Cmnd. 247 (HMSO, London, 1957).

18. M. Warnock, *The Times*, 30 May 1985, p. 12.

19. Warnock Report, paras 8.10ff.

20. I. Kant, *Grundlegung zur Metaphysik der Sitten*, 2nd edn, p. 67, translated as *The moral law* by H. J. Paton (Hutchinson, London, 1948), p. 96. For Kant's meaning, see Paton, p. 98.

21. *The ethics of clinical research on children*, report of working party, chairman G. Dunstan (Oxford University Press, Oxford, 1986).

22. M. Tooley, 'Abortion and infanticide', *Philosophy and Public Affairs*, vol. 2 (1972), p. 60.

Part II
The Perfect Baby

6

The Perfect Baby: Introduction

Ruth F. Chadwick

For a very long time human beings have rejected defective babies. The Spartans used exposure for those who were not up to scratch; Plato in the *Republic* recommended that babies whose conception and birth did not fit the prescribed guidelines should be hidden away in a dark and secret place.[1]

In more modern times well-publicised cases such as the Arthur trial in the UK and the Baby Doe case in the US have made it clear that parents still reject babies with severe defects, but these days the practice of infanticide is regarded by many as morally wrong, despite attempts to show either that it is not, in principle, morally different from abortion, or that the sanctity of life doctrine on which objections to infanticide rest is unsound.[2]

However that may be, abortion for handicapped foetuses is a very common practice. The 1967 Abortion Act lists handicap of the foetus as one of the grounds which justify abortion, and amniocentesis is routinely offered to women thought to be at risk of carrying an abnormal foetus. Although, of course, abortion is itself by no means uncontroversial, the fact of handicap is generally regarded as one of the best reasons for having one. There is also a considerable amount of public sympathy with the parents and doctors involved in cases where babies are allowed to die, as in the Arthur case.

In these modern cases it is the plight of those most closely involved in the drama that most commonly receives attention, especially from the media, but we should not forget that Plato was interested in the quality of babies from the point of view of the interests of the state. This strand of thought has not disappeared. Thus Mason and McCall Smith mention the 'quite proper interests of the State in reducing the incidence of genetic disease'.[3]

The state has at times intervened, and drastically, with such policies as compulsory sterilisation. In the notorious US case of *Buck* v. *Bell*, concerning an application for a sterilisation order for Carrie Buck, the judge remarked: 'Three generations of imbeciles are enough.'[4]

Thus in addition to the disposal of those conceived or born with genetic disease, steps have been taken at certain times and in certain places to prevent the conception of handicapped people.

The modern practice of genetic counselling, far less drastic than compulsory sterilisation, also presupposes that it is undesirable, from some point of view, to bring into existence handicapped people if it can be avoided.

Similarly, the Warnock Report recommends that no embryo which had been experimented upon should be returned to a woman.[5] Why not? Presumably because it might be damaged. What other reason could there be? Warnock also recommends that, although children born 'by donation' should not be allowed to know the identity of the donor, they should be permitted to have access to information about the donor's genetic health. This shows how important genetic health is considered to be. For when that child born by donation grows up he or she in turn may want genetic counselling. So while Warnock does not give detailed consideration to genetic engineering, beyond saying that possibilities are speculative,[6] the Committee is concerned about the genetic health of children.

Thus we have to consider, not only the desire to have a child, but the desire to have a child of a certain *quality*. As Jeremy Rifkin says, 'The fact is we humans want perfect babies.'[7] The first thing a mother wants to know after the birth of her child is whether it is all right, whether it has ten toes and ten fingers, etc. But if the desire to reproduce is as strong as it is made out to be why are we so concerned about quality?

The desire for quality

The parents and family

There is no denying that a handicapped child may be a terrible burden for parents, and take their attention away from other children, who may feel their own experience to be impoverished thereby.

On the other hand we are familiar with the moral growth that people experience in coping with crises of this sort. Errol E. Harris suggests that we need struggle precisely to provide us with moral exercise. He argues that many of the higher values are constituted in large measure by difficulty of achievement and that it is fallacious to assume that making things easier for people will lead to a better state of affairs.[8]

But even those parents who are willing to make enormous sacrifices to care for severely handicapped children worry about what will happen to those children when they themselves no longer have the health and strength to look after them. And surely life provides sufficient morally testing experiences without requiring us to have handicapped others to respond to?

Society

The costs to society of the existence of handicapped babies are generally interpreted in terms of economics. It costs money, uses up resources, to supply back-up or replacement care for families which need support. Some may feel that they are no longer willing to provide those resources when it is perfectly possible to avoid bringing the handicapped into the world. Thus Amitai Etzioni envisages propaganda to the effect that people should not be allowed to 'dump' defective babies on society.[9]

The gene pool

Next we come to the gene pool. It is argued by those who have an interest in eugenics that by continuing to produce babies with genetic defects, we are adding to the 'genetic load' that is carried by the species. By not exercising quality control over the species now, we are not only bringing into the world individuals who have a low probability of a satisfying quality of life, but we are also storing up trouble for future generations, who will be faced with the results of our lack of forethought.[10]

These are arguments of a broadly consequentialist sort, pointing to undesirable consequences for parents, society and the species. But what about the child produced?

Wrongful life

A first thought might be that a child brought into existence with a genetic handicap, when this could have been avoided, has been wronged. However, English law has discouraged 'wrongful life' suits, partly on the ground that there cannot be a duty to take life.[11]

However, this cannot be the whole story. Whatever the courts have held, there is still the moral question, about whether a wrong has been done when a handicapped person has been brought into existence.

From one point of view it seems that it is a nonsense to say that a person has been harmed, because at the time the relevant action was performed or the relevant decision taken, there was no one in existence to harm. Further, at least in most cases where the question has arisen, there was no choice to be made between producing X with a handicap or X without handicap, but between producing X with a handicap or no X at all.

If one looks at the situation from a consequentialist point of view, however, if X alive is so miserable that he wishes he were dead, and X's existence does not create happiness overall sufficient to outweigh this misery, then it was wrong to bring X into existence. This avoids the difficulties in saying that X has been harmed, but it also looks at the question from an impersonal point of view, and it may seem unpalatable to some that a calculation is made of the total benefits and costs, rather than concentrating on poor X's fate.

Just thinking about X, then, from the consequentialist point of view, we can say that the more handicapped he is the lower probability he has of achieving a high quality of life, interpreted in terms of preference-satisfaction. Therefore anybody who is concerned about promoting quality of life has a reason to bring it about that a child is produced without handicap.

At this point the objection will be made that many handicapped people are able to achieve a very happy life with severe handicaps, whereas many apparently 'normal' people are miserable. This may be true; the connection between handicap and quality of life may be a contingent one only. But that does not alter the fact that most of us would choose not to have a handicap if we had such a choice, because the probability is that we shall have more choices open to us without the handicap. So why should we not also choose that for our children?

Hence there are all sorts of reasons why it might be thought wrong to bring into existence a baby with a genetic defect: if we care about the children we produce surely we want to give them the best possible chance of achieving a satisfying life. It is consistent with the child-centred approach advocated in Part I that we give weight to children's welfare.

Further, in addition to the moral question about the effect on the individual, family and society, parents have prudential reasons to want to avoid handicap, as does government. Finally, there is the eugenic aim of cleaning up the gene pool.

We have spoken in terms of there being the possibility of choosing to avoid handicapped children. The new reproductive technologies have added a new dimension to these questions. They may offer the promise of greater control over, not only the ability to have children, but also the *quality* of these children. In fact we have already seen the transformation of AID from a technique for remedying infertility to a eugenic scheme in Robert Graham's Repository for Germinal Choice in Escondido, California, where women have been inseminated with sperm from Nobel prizewinners in an attempt to produce superbabies.[12]

The papers in this section deal with aspects of the issue of controlling the *kind* of children we have. The questions to be addressed are, to what extent is it legitimate to attempt to do this, and in what ways does the advent of the new technology affect the practicability of the enterprise?

We have been talking so far in terms of genetic control, but it is envisaged that the new technology will also give us increased potential to control the sex of future children. These problems must be considered in turn.

Genetic problems

Early in the twentieth century it was established that chromosomes are the carriers of genetic material. It proved difficult, however, to discover what in the chromosomes was the genetic material itself. It was not until the 1950s that DNA came to be recognised as fulfilling that role. In that decade much was learned about the structure of DNA (deoxyribonucleic acid). Experiments suggested that DNA, while made up of four nucleotide bases, adenine, guanine, cytosine and thymine, was a long chain formed out of these four in many different sequences.[13] The innumerable possibilities of sequences accounted for the variety that the genetic material has shown itself capable of producing.

In 1953 Watson and Crick made their proposal that DNA's structure was a double helix, i.e. a molecule composed of two coiled strands.[14] It is commonly compared to a spiral staircase. When DNA replication occurs as cells divide and multiply, the two sides of the staircase separate and the nucleotide bases are stranded without partners. In this situation they attract 'free' complementary bases, so that eventually two new complete spiral staircases are formed. This is how DNA reproduces itself.

Where in the DNA do we find a gene? As early as 1909 Garrod[15] had shown that the human disease alkaptonuria, which appeared to be inherited as if due to a single recessive gene, was caused by the lack of some enzyme. It seemed that there was some connection between genes and enzymes. Later experiments indicated that enzymes were not the only proteins to be controlled by genes.

Proteins consist of amino acids. Work by Ingram[16] and by Yanofsky[17] and others provided evidence that the gene should be defined as a section of DNA, a sequence of nucleotides, that embodies a code of instructions for amino acids. Crick and others discovered in a series of experiments in 1961 that the nucleotides are to be read in a triplet code.[18]

A single copying error or mutation can have disastrous effects. For example, in the normal individual, the beta-chain of haemoglobin contains an amino-acid sequence

val — his — leu — thr — pro — glu — glu — lys

A genetic mutation produces the following sequence in those who are homozygous for the gene:

val — his — leu — thr — pro — val — glu — lys

This individual has the genetic disease sickle-cell anaemia.

Different forms of the same gene existing in a population are known as alleles. As chromosomes occur in pairs, an individual may have two allelic forms of a particular gene. If that is the case, he is said to be heterozygous for the gene. If he has the same allele, he is said to be homozygous.

Genetic disorders may be divided into the following kinds:

(1) Dominant disorders
A dominant disorder is one whose effects will be felt even if the individual is only heterozygous for the gene. An example is Huntington's chorea.

(2) Recessive disorders
Recessive disorders manifest themselves in the phenotype (which is the individual as he appears) only in the homozygous state, i.e. if the gene is inherited from both parents.

(3) Sex-linked disorders
One pair of chromosomes is associated with sex differences. A female has two X chromosomes and a male has one X and one Y. The X is much larger than the Y. If a woman has a recessive gene on one chromosome she will not be affected, because she also has a normal allele. But if a man has the gene on his X chromosome he has no normal alternative, and will be affected. Haemophilia is the classic example of this.

In addition to these genetic defects problems are also caused by

(4) Chromosome aberrations
Some chromosome aberrations are connected with the number of chromosomes, others with the structure. The latter kind arise out of the breaking of chromosomes, which may cause the loss of, or rearrangement of, genetic material. Chromosome aberrations usually have very serious effects and are responsible for many spontaneous abortions.

Among those which involve a departure from the norm as regards the number of chromosomes, the familiar Down's syndrome results from the presence in triplicate of chromosome 21. Another aberration of number is the XYY syndrome. Males who have this extra Y have certain abnormal characteristics including unusually tall stature, and attempts have been made to link the syndrome with a tendency to violent behaviour.[19]

So these are the types of genetic defects and chromosome abnormalities which constitute the genetic load which the species carries. We have suggested that when measures are available to ensure that children with defects are not produced, parents and governments may want to avail themselves of them. What are the possibilities?

Genetic solutions

Eugenic breeding

One of the oldest solutions, and one supported by Francis Galton, is that those who are genetically weak should simply be discouraged from reproducing. Eugenists have varied on whether they support merely a programme of incentives or compulsory measures such as state-controlled breeding and compulsory sterilisation.

How effective are such programmes?

As regards dominant mutations, many of them are lethal, and so those who suffer from them die young. Thus they are unlikely to pass on their genes anyway. Some dominant disorders arise from fresh mutations, about which nothing can be done by eugenic breeding.

However there are those, like Huntington's chorea, which do not make their appearance until later in life, when the individual concerned may already have had children. This is one defect that is passed on by affected parents. Approximately half the children will have the disease. A way to avoid this would be to identify the carriers and prevent their reproducing.

The appearance of recessives in the phenotype could be prevented if carriers of the same recessive did not have children. When two people who carry the same recessive reproduce they run a one in four risk of having an affected child, and approximately half their children will also be carriers. Again, control relies on being able to identify carriers, by screening procedures.

The same applies to sex-linked disorders. Appearance of these in the phenotype could be prevented if carriers could be detected and if they refrained from reproducing.

However, while eugenic breeding could have some impact on genetic defects, it would have very little impact on chromosome aberrations. Unlike genetic mutation, they do not follow an inheritance pattern, for the aberration may arise in the gamete of a parent and be transmitted to the offspring, but no further. Such aberrations have a greater probability of occurring with greater parental age, so some eugenic effect can be achieved by early parenthood, but other techniques have been suggested for detecting and eliminating these conditions.

Foetal screening and abortion

It has for some time been possible to screen the foetus by, for example, amniocentesis. In this process a needle is inserted through the uterine wall so that the amniotic fluid can be withdrawn and the cells examined. Then if the foetus is found to have e.g. a chromosome aberration, it can be aborted. Earlier and better methods of screening are constantly being researched.[20]

Gamete donation

Warnock says:

> Even though [couples] do not require the donation as a treatment for infertility, it seems right to us that . . . they should be offered, as part of the process of genetic counselling, the facility to use a technique which will prevent handicap in the next generation. We know that, where the hereditary condition may be transmitted by the male, AID is already quite often suggested.[21]

Thus the technique of AID has implications for genetic quality in so far as donors, and their sperm, can be screened. This is one way in which the advent of reproductive technology can make an impact on genetic control. Similar considerations will apply to egg donation.

Gene therapy

Eugenic breeding, abortion and AID are ways in which certain defects could be prevented from appearing in the phenotype, but in recent years it has been suggested that it might be possible to treat genetic defects at the genetic level. This would be an alternative to simply trying to eliminate those who suffer from them. Here we are talking about gene therapy, using techniques of genetic engineering.

Gene therapy is described in the papers by Robert Sinsheimer and W. French Anderson. The aim is to insert a gene into the body of a patient who lacks a gene or has a faulty gene. There are, as Anderson points out, serious practical difficulties to be overcome. Even if the new gene is introduced, how can we be sure that it will be properly controlled in its output? In principle, however, the possibility is there for curing genetic diseases at the genetic level.

Anderson also discusses the different possibilities of somatic or germ line therapy. If the therapy affects only the individual's somatic, i.e. body cells, he would still pass the defect on to his children who would also need gene therapy. But if the germ line is also affected then one is not only treating the patient but affecting future generations. As Anderson points out, this adds a further dimension to the ethical issues.

Cloning

One method of cloning is to divide the embryo into two or more parts which will develop separately. These members of a clone would all have identical genotypes. Jerome Lejeune in his paper describes how such a process of twinning might be carried out so that one of a pair could be sacrificed to provide a check on the quality of the other.

The more popular, science fiction sense in which cloning is understood is the suggestion that identical copies of adults could be produced via the process of nucleus substitution, i.e. taking the nucleus with its genetic information from the body cell of an adult, placing it in an enucleated fertilised egg, and allowing it to develop. Thus if one partner of a couple was in danger of passing on a defect, that couple could have a child by cloning the other partner. But some feel that this *is* science fiction as far as humans are concerned.[22]

In vitro fertilisation

In vitro fertilisation provides the opportunity for screening of embryos, e.g. by twinning as described. Robert Edwards for one has suggested that defective embryos could simply be disposed of at that stage.[23] On the other hand it might provide the opportunity for manipulation of the embryos.

These are the methods suggested for solving the genetic problems. However, the possibilities do not come to an end at this point. It has been noticed that in addition to removing defects from the species, some genetic policies could also be employed to promote *improvements* in genetic quality, so that the human species, or individual members of it, could attain new heights.

Some of the methods described could be used for this purpose. Breeding programmes, for example, could encourage those who are outstanding in some way to have children. This is known as 'positive' eugenics, as opposed to 'negative', which aims to eliminate genetic defects.

Perhaps, also, it would be possible to introduce favourable genes into individuals, e.g. to make them taller than average, rather than simply giving those genes the absence of which cause genetic disease. This is what Anderson calls 'enhancement genetic engineering'. He makes a distinction between this and eugenic genetic engineering, which he sees as the attempt to manipulate complex traits such as intelligence and personality.

AID already has been used in a positive eugenic way, as we have pointed out. IVF may offer the possibility of manipulating embryos, but Chris Graham in Jonathan Glover's 'Horizon' programme,[24] suggested that it would be simpler to get the results by cloning rather than by squirting genes into eggs.

Moral problems

The popular conception of eugenics and genetic engineering is an alarming one to many. Eugenics has been associated with 'master race' theories and Nazi policies. Genetic engineering has been presented in a disagreeable light by some science fiction writers. In David Brin's *Startide rising*, 'uplift' is seen to go badly wrong on some of the dolphins engineered. Further, the attention of such writers as Ira Levin has centred on cloning; in *The boys from Brazil* he depicts an ex-Nazi who has brought into existence ninety-four copies of Hitler.

But these sensational ideas are far from being the whole story. Opinions differ both as to the likelihood of various practices being attempted and as to the possibilities of their success. There are also great differences of opinion as to the moral issues.

So let us look at some of the objections to genetic intervention in man.

Liberty

First of all, policies such as state-controlled eugenic breeding arouse objections on the grounds of liberty. Most of us would presumably prefer that our choice of mate was not made by the state. There is scarcely room for doubt that a preference-utilitarian would have to take such a preference seriously.

But these days few would advocate so crude a scheme as that envisaged by Plato in Book V of the *Republic*, where the citizens are to be deceived as to what is going on in the manipulation of the quality of the stock of the human 'herd'.

However, countries which have had compulsory sterilisation laws have tried to liken transmission of deleterious genes to the transmission of other agents dangerous to health, and to liken sterilisation to vaccination,[25] which is now widely accepted despite the controversy over whooping cough.

How safe is the parallel? In the first place, we all carry dangerous recessives, so if we were to ensure that no deleterious genes are transmitted then everyone should be sterilised.

Secondly, in the vaccination case it is likely that the person vaccinated will also benefit by receiving immunity to the disease. The person compulsorily sterilised has no such benefit to which to look forward if the sterilisation will not satisfy a desire she has. The two cases are therefore very dissimilar.

Compulsory screening may not seem quite so drastic. It does not lead in itself to forbidding people to have children, for example. It may lead to persuasion and incentives of course.

But even persuasion is not unproblematic. C. K. Chan's paper describes a eugenic scheme recently proposed by the government of Singapore, which relied on incentives, largely to encourage adults thought to be 'desirable' from a genetic point of view to have children. Does the replacement of compulsion by incentives make it acceptable? This is by no means obvious. Etzioni imagines a propaganda campaign using such advertisements as 'You Must No Longer Bring a Mongoloid Into the World.'[26] It is far from clear that this is satisfactory as a less restrictive alternative. Pressure of public opinion can have extremely unpleasant side effects. At the present time many people experience shame and guilt for genetic disorders in their offspring, and this may only be increased by a persuasion campaign. For however careful people are, some defective individuals will slip through the screening net and disorders will be produced.

Geneticism and elitism

A further objection put forward is that eugenic schemes are elitist. This is a problem that particularly concerns C. K. Chan.

The very prospect of eugenics seems to imply that some people are better than others. People naturally resent the suggestion that there may be anything wrong or inferior about their genes. For what one's genetic constitution is plays an enormous part in *who one is*. (Hence the concern with genetic history in Part I.) It is feared that 'geneticism' may lead to discrimination.

Thus there arises the demand that people should be accepted as they are, and that if changes are to be made, they should be made in the environment, which should be adapted to people's needs, whatever genetic flaws they may have. Geneticism, it is argued, will lead to even less concern about the environment than there is now among those in positions of power.

There is a certain amount of evidence to support this view. Thus Tabitha Powledge in 1976 described the practice of General Motors' lead battery plants as being to insist on the sterilisation of those women of child-bearing age who wished employment. Lead has deleterious foetal effects, and to avoid these the company preferred a policy of sterilising female employees to cleaning up the lead oxide.[27]

While this is a fairly stark example, it would be wrong to see the choice between changing the environment and changing the gene pool as totally opposed alternatives.[28] It is not clear that if one adopts a policy of genetic engineering or eugenics, it must be for the benefit of powerful interests in society rather than for the benefit of individuals. We have argued that genetic health may be of instrumental value in the individual's achievement of a satisfying life.

Nevertheless it has been argued that a concern for genetic control is inevitably discriminatory.

Marc Lappé has suggested that a concern for genetics will affect the way in which we regard and behave towards persons: 'Where genetic models supplant environmental or nutritional ones the danger is that victims of socially related diseases will be held responsible for their fate and the rationale for remedial programs subverted.'[29] In other words, if we think that problems are genetic we may give up trying to improve the environment for people, when in fact their problems may have an environmental cause. We have seen how this has happened in the cases of race and sex.

Those who have held that e.g. blacks are genetically inferior to whites and that women are genetically inferior to men, in some cases also hold that it is useless to try to establish equality.

Screening programmes, it is thought, may lead to a situation in which, when some genetic abnormality is found, the 'patient' is held to be sick irrespective of his symptoms. This has happened with the sickle-cell trait. Sickle-cell anaemia is a genetic disease which affects those who are homozygous for the sickle-cell gene. Those who inherit the gene from only one parent do not have the disease but their red blood corpuscles may become sickle-shaped. This is the sickle-cell trait. The heterozygous condition confers the advantage of immunity to malaria. But in the US persons with sickle-cell trait have been discriminated against in the mistaken belief that they had the disease.[30]

Further, in some cases where certain males have been found to have the XYY chromosomal complement, this fact has been held to be grounds for continued imprisonment. And this was despite evidence that the defect to be remedied was not XYY but societal, because the rate of chromosomal non-disjunction leading to the XYY karyotype increases among the lower socio-economic groups, and may be due to nutritional deprivation.[31]

Discrimination on the basis of genetic traits *irrespective* of phenotype symptom is as unjustifiable as discrimination on the basis of sex or colour. Lappé has therefore suggested that in deciding whether to adopt genetic models we are making a moral decision. When we adopt a genetic model we must suspend alternative hypotheses, and so the question we must ask is: 'Are these models appropriate for formulating social policy?'[32]

Lappé seems to have confused two distinct questions here: 'Are the genetic models correct?' and 'How should we formulate social policy?' We cannot say that the genetic models are false simply because we do not like the implications. Lappé thinks that the genetic models carry with them a 'highly charged component of determination'. Thus for example if we decide to treat IQ as a genetic problem then we are threatening the moral basis for ensuring equality of opportunity.[33]

The important thing is that we should find out whether or not traits are genetically determined, for it is only by knowing more about this that we shall know how to help individuals who are genetically disadvantaged. Knowledge can lead, not to discrimination, but rather to compensation of those who are in need of it. As Hook has pointed out, evidence for genetic influence is not evidence for genetic fatalism.[34] His view is that genetic studies can reveal who is in need of what environmental help. But we may be able to go further than that and say that genetic studies can reveal both who is in need of what environmental help and who is in need of what *genetic* help.

After all, Lappé himself, in criticising the use made of genetic models, uses examples of where the bad results arose from ignorance. The XYY syndrome, according to his own analysis, may have been misused because of lack of, rather than the possession of, knowledge.

However, in some circumstances our idea of what 'help' is appropriate may be resented. It has been suggested that if it is possible to find out that certain people are genetically predisposed to certain conditions, this 'genetic prophecy' will enable us to make decisions about what work a person would be suited to. Thus if a person has a genetic predisposition to certain forms of cancer this may cut him out of work with certain chemicals which may be carcinogenic. Of course, he may not want to work with carcinogens anyway, but he may appreciate the chance to make up his own mind about whether to take the risk.[35]

In Robert Silverberg's novel *Master of life and death*, Walton, who is in control of the euthanasia 'Happysleep' programme, breaks the law to save a child who is diagnosed as being potentially tubercular. At the end of the book he meets father and son and is surprised at how healthy the boy looks. The father points out: 'That's all it was; potential.'

Discriminatory means

In addition to the fear that geneticism itself leads to discrimination, however, we must consider to what extent various means of genetic control involve discrimination.

Breeding

The methods of the 'old' eugenics, where some adults were thought to be suitable for parenthood while others were not, arouse complaints of elitism and preferential treatment for certain groups in society. Plato realised that this would not go down too well with those who were not to be allowed to reproduce, so he proposed to keep the whole plan secret.

C. K. Chan, describing Prime Minister Lee Kuan Yew's

schemes to produce a high-quality population in Singapore, ascribes them to the 'ideological expression of privileged class interest'.

Policies which depend upon one set of people taking decisions about which, of another set of people, should be allowed to breed, or which claim that e.g. 'educated women' have a 'duty to breed',[36] and set about offering incentives to see that they do, arouse resentment. The Singapore plan included incentives both to 'superior' individuals to have children, and to others to refrain, the latter being dressed up as an anti-poverty measure.

Such policies involve discrimination which is held to be unjustified discrimination, because a judgement is being made about superiority and inferiority in human beings, not with relation to their capability for e.g. a job, but as regards their worthiness to reproduce. It involves saying to certain adults 'We do not want any more of your kind.'

The advocates of such proposals may say that what they intend is not disparagement of certain sorts of human being overall, but a judgement that some people have superior credentials for becoming genetic parents than others, and that this is an activity for which there are criteria of suitability, just like many other activities.

Objections may take several forms.

First, there may be an objection to the particular criterion chosen. Discrimination is held to be unjustifiable where a characteristic is made relevant in a context where it is irrelevant.

Looking at the Singapore plan, where intelligence was the chosen characteristic, the objection could be that it was based on bad genetics. Clever parents do not necessarily have clever children, and thus to attempt to build a eugenic policy around this is to court disaster.

Further, where a privileged group is making these decisions, its members presumably think that they have the relevant characteristic. But they may have been successful in the phenotype for all sorts of reasons which may not reflect their genetic inheritance.

A related objection is that offering incentives to those seen to be among the privileged, will be seen as widening the gap that already exists between them and everyone else, although, ironically, if they increase their number by too much the notion of privilege will itself cease to have meaning. It is noticeable that Plato did not want to raise the standard of the whole population, but only of the Guardian class.

A deeper objection to eugenic proposals of this type would be not that the criterion is mistaken, or that it depends on reinforcing the privileges of contemporary society, but that there are *no* criteria by which some people may be judged superior for the purposes of reproducing themselves. Hence the objections to schemes like Hugh Lafollette's suggestion of a licence to reproduce, although he was thinking in terms primarily of preventing violence to children.[37]

Is it really the case that there are no criteria which would justify saying that one group of people should be encouraged to reproduce and another set discouraged? Are all decisions on this matter reflective of an unjustifiable elitism?

Counselling

When we consider proposals which involve a degree of coercion, or the kind of incentives that are hard to refuse, other considerations can easily cloud the issue, so let us consider genetic counselling, where no coercion is involved, and there are no material rewards offered either to encourage or discourage prospective parents. Can genetic counselling be carried out without unjustifiably discriminating for or against kinds of human being?

Where we are talking about trying to prevent genetic disease, it may seem as though the discrimination can be removed to another level. For it could be argued that we are trying to diminish the frequency of certain *genes*, not certain kinds of people.

Success in diverting the effects of discrimination on to the genes will depend on the extent to which individuals feel their identity to be bound up with their genotype. In the circumstances in which counselling for genetic disease would be most helpful, it may appear difficult to understand why the extent should be great. For in testing a couple and finding that they both carry a dangerous recessive, to advise them of the risks of producing a child homozygous for the gene would not be to say 'We don't want any more people like you.' After all, their recessives may have no effect on their phenotype.

There are, however, exceptions to this, as in the case of sickle-cell trait. Because this trait tends to be associated with certain races, screening has had unfortunate consequences. Not only have those who were diagnosed as having the trait been treated as though they were sick, but screening programmes have provoked race riots in parts of the USA.[38]

But this, like the consequences of geneticism which worried

Marc Lappé, is the result of limited understanding. The answer is surely more, not less, knowledge of genetics. C. K. Chan also makes the point in relation to the Singapore scheme that the plan is based on unestablished theories taken as fact.

So counselling voluntary abstinence from parenthood to decrease the frequency of genes may avoid the charge of unjustifiable genetic discrimination if it is understood that the genes in question are not in any way related to the performance of the phenotype of the adults counselled.

However, such counselling programmes may arouse suspicion about the motives of the funding body and the decisions they have taken as to what characteristics should be labelled 'undesirable'. The definition of what constitutes a genetic disease or defect may not be as objective as we should hope. However, the counselling process may be more palatable if the argument is presented in terms of the quality of life of the resulting children, rather than the good of society, as in Singapore.

Abortion and embryo destruction

When we consider the elimination of genetic defects, not by preventing conception, but by *killing* those who suffer from them, we enter upon a new set of problems.[39]

Are we suggesting that the handicapped have less right to live, with consequent repercussions for the handicapped who are presently alive?

Let us consider parents who discover that their foetus has a genetic abnormality, and decide to abort it. Is this an instance of unjustified discrimination against the handicapped?

It is conceivable that genetically handicapped people may feel belittled by this decision, on the grounds that the more commonly this sort of decision is taken, the less tolerated they will be in society. Parents in this position may also find it more difficult to choose not to have an abortion.

So although the discriminatory force of eugenic abortion is directed against foetuses, and may therefore seem less serious, e.g. because the foetus will not live to feel resentment, nevertheless the same decision is being made about superior and inferior human beings, and those who have slipped through the screening net may feel the effects of this.

However, the important thing to remember here is that it is not only *handicapped* foetuses that are aborted. Thousands of healthy foetuses are also aborted, for example if they constitute a threat to the health or welfare of the mother. But this does not lead to the view that any adult who is a threat to someone's health or welfare is less valuable or less worthy of respect. The important boundary here (at least from a psychological point of view) is that between foetuses and adults, not between handicapped and healthy. Those who fear a psychological slippery slope have to admit that we seem to be able to put these abortions into another category and not let them affect how we respond towards other living humans. (Newborn infants are a special case.[40]) As long as we maintain this boundary between foetal life and other life, there is little reason to think that eugenic abortion will lead to less respect for those handicapped people who are alive. If the adult handicapped were thought to be ideal candidates for euthanasia, that might be a different matter.

Thus when Elizabeth Bouvia, a victim of cerebral palsy, argued in court that she had a right to starve to death, groups representing the rights of the disabled demonstrated outside, arguing that she had no such right because of the side effects that would result for the disabled. They felt that less effort would be made to provide facilities for the disabled if she was permitted to relieve society and herself of what she felt to be the burden of her life.[41]

But the thinking behind eugenic abortion is not necessarily that genetically handicapped people are less valuable in some overall sense than others. The idea may be that foetuses are not yet people and we are still in some sense deciding what sort of a child to have.

To some these views will appear not only misguided but morally wrong, as they take the view that foetuses *are* people and that eugenic abortion does imply that the genetically handicapped are less valuable than other people, but what has been suggested here is that the real difference of opinion is over whether foetal life is less valuable than other life, not whether the handicapped are less valuable.

When we turn to embryos, the situation is akin to foetuses but at an earlier stage. Thus if eugenic abortion does not affect the way we value other humans, then destruction of handicapped embryos should not either.

AID

Some of the reproductive technologies give the possibility of controlling the genes a future person will have. Thus if we consider AID, the donors are screened for their genetic health, and it would seem wrong for a practitioner knowingly to use sperm from a donor whose genetic health was unsatisfactory.

But it would seem odd to categorise this decision as elitist, whereas Robert Graham's Repository *has* been so categorised. This may be because in the latter case the donors chosen have been selected on the basis of their success in the phenotype without any knowledge of the genetic basis for this.

However much people may overcome their handicaps and achieve moral worth by so doing, it would seem very strange to *choose* to have a handicapped child, where there was a choice. This is because if we want the best for our children, we will choose to maximise their chances of having a high quality of life, and there seems to be some link between their genes and the quality of life we expect them to be able to achieve, however mistaken we may turn out to be in our predictions.

If we say that it is better, all things considered, to produce a child who is not handicapped rather than one who is, this does not commit us to saying that handicapped children are less worthy of respect. It is just that, other things being equal, it is preferable for the individual, the family, and the society.

After all, if it were possible to cure a handicap by a drug which was known to have absolutely no undesirable side effects, we would. So trying to eliminate genetic problems by e.g. gene therapy, if safe, would not entail the view that handicapped people are less worthy of respect, even though curing them is, in a sense, eliminating them.

There are obvious utilitarian reasons for opposing elitist discrimination, on the grounds that it causes resentment among those discriminated against and causes tensions in society.

But the AID case shows clearly that choices about what is better and worse, as regards genes, need not involve elitism or unjustifiable discrimination. Nor is it clouded by the issues of either inducements or killing.

But it may be held that it is clouded by another issue, viz. that of deliberate as opposed to natural creation. For example there may be thought to be a difference between accepting God's will in the existence of a genetically handicapped foetus, and deliberately creating one.

Artificiality

The practitioners of reproductive technology have insisted that they are not creating life, but only assisting in its creation. In IVF, for example, all they can do is bring the sperm and egg together in the dish. They cannot create the sperm and egg. However, although they cannot create life, they may eventually have considerable control over what the life they assist to create is like. Is there anything wrong with this, in principle?

To what extent it may be possible is a matter for speculation but let us consider Caryl Rivers's imaginary man programmed to be legless, apparently a useful trait for an astronaut.[42] Certainly people are born legless in nature, but to design a man to be legless seems peculiarly unpleasant. However, if he were designed to be a complete man it would not seem quite so distasteful, so it seems our objections to it cannot rest on the fact of design alone.

Plato, in the *Republic*, put forward the view that each man should perform the task for which he is most fitted.[43] Even those who agree with this must admit that this is very different from suggesting that each person should have a pre-ordained task and then be cut out for it. Suppose this person does not *want* to be an astronaut. In order to ensure that everything will go smoothly one would have to go one step further and 'adjust' his thoughts and desires as well, a possibility referred to in Part I.

Our objections to this could rest either on the view that we are restricting this person's choices, or that we have some minimum acceptable level of welfare, and being legless falls short of this. An individual may not necessarily be worse off for having an artificially designed genetic constitution, and so not all human design can be ruled out on these grounds, but we need to be confident that change will not have undesirable consequences.

It seems to be the case, then, that if we are to have deliberate creation (or assisted creation) the persons so produced must be at least as complete, genetically speaking, as people generally are.

IVF practitioners implicitly accept this when they suggest that risks of abnormalities are no greater than those that are attendant upon normal methods of birth.[44] It is not the artificiality that is the important thing, but the consequences for the people produced.

However, we have argued that when something is described as unnatural, it may not in itself be a satisfactory argument but it acts as a counsel to be very sure that the benefits outweigh the consequences, especially when we are talking about what is natural to our kind.

If we turn from thinking about the individuals to thinking about the species, we can see that there is little reason to think that artificial selection is in itself necessarily inferior to natural. As Passmore points out, natural selection has no special interest in man; artificial selection can have.[45] If the environment underwent a radical change, and the human species failed to adapt, the species could die out. Artificial selectors could try to foresee this and design or plan for a future type that could ensure the survival of the species, as in Olaf Stapledon's *Last and first men.*

In a sense it is impossible to replace natural selection by artificial selection. Some have suggested that the statement of the theory of natural selection is itself analytic in that it states that those who are fittest will survive, where fitness is defined in terms of survival.[46] Artificial selection is an attempt to make the differential work to the advantage of those whom we want to survive.

Risk

We have suggested that if we could cure genetic problems by a safe procedure, we would do so. But is there a safe procedure? The main moral problem in connection with the possibility of gene therapy, as brought out in W. French Anderson's paper, focuses on the risks of harm involved.

There are many things that may go wrong in the course of eugenics and genetic engineering. There may of course be unintended errors, and the results may not be what we wished. In the case of gene therapy applied to individuals, there is the risk that it may not work, as with any kind of medical intervention.

But Anderson wants to draw a distinction between somatic and germ line therapy, in terms of the numbers of people who may be put at risk of harm. His point is that somatic cell therapy will introduce a new gene into the body of the patient, but will not affect reproductive cells, so it will affect only the particular individual on whom the therapy is carried out. Germ line therapy, on the other hand, will affect the gametes as well, and thus future offspring.

Anderson sees somatic cell therapy as being in principle no different from any other experimental treatment. In this he agrees with Arno G. Motulsky, who says that it is 'conceptually no different from any therapy in medicine'. Motulsky takes the view that it can be seen as 'euphenics' rather than eugenics, as it does not actually enable that individual to transmit the new gene.[47]

Germ line therapy is thought to introduce a qualitative difference, because the effects will be perpetuated in future generations, and society as a whole will be affected rather than the individual patient.

Thus Anderson sees a major difference in that while an individual patient can choose whether or not to accept the risks of harm associated with somatic cell therapy, germ line therapy may impose risks on people who have had no say in the matter.

Here we have a distinction between voluntary and involuntary risks. It is fashionable to speak of an 'acceptable level' of risk. But does this apply to both the voluntary and the involuntary kind?

In the case of a voluntary risk it may seem that we have a simple test of whether or not the risk is acceptable: does the person concerned find it acceptable? I am going to ignore for these purposes the question of how voluntary any choice can really be said to be. It seems clear that someone who chooses to be a motor racing driver in full knowledge of the statistics of risks of death in some sense voluntarily accepts those risks.

Is it wrong to expose people to involuntary risks? Of course everyone is subjected to involuntary risks in any case, such as the risk of catching influenza. But it could plausibly be argued that one section of the community, e.g. scientists, should not deliberately subject others to such risks.

How do we determine whether an involuntary risk is acceptable? An approach attempted by Lord Rothschild was to establish an index of risks[48] on the basis of what actually does cause anxiety. But this seems unsatisfactory because if what were most important were actual anxiety, we could reduce this by keeping people ignorant of the facts. People may have a preference for not suffering anxiety, but it seems plausible that they also have a preference for not having a good reason to feel anxious in the first place.

Given that everyone is exposed to involuntary risks anyway, it may be justifiable to introduce a measure which also involves an involuntary risk, but one that is less than the first. For example, if every infant runs the risk of catching a disease which could be fatal, one plan is to vaccinate all children against it, a policy which would also expose them to the risk of death, but one statistically less than the first.

So there may be circumstances where it is justifiable to expose people to an involuntary risk. It is necessary to determine if this counts as one of them.

What are the possible harms and benefits?

If health is seen as a good, whether because people have a preference for it or for any other reason, then *a fortiori* genetic health is a good that we have an obligation to pursue. Anderson sees gene therapy as part of the effort to alleviate human suffering. Health is valued as a means to the achievement of a satisfactory quality of life. Those who do not subscribe to a punishment theory of disease must hold that we should think it desirable to try to cure diseases where possible.

Carrying out a cost-benefit analysis for somatic cell therapy seems a fairly straightforward thing, in principle. But with germ line therapy, it is more difficult to get an idea of what the possible harms and benefits are, and how probable they are. In the first place germ line therapy involves severe practical difficulties.

More seriously, it is difficult to appreciate exactly who might be benefited or harmed and in what way. Perhaps this is due to a failure of the imagination. Let us take benefits first. Suppose we have the possibility of benefiting both an embryo and its future off-spring by eliminating a genetic handicap. As far as the embryo is concerned, we can understand at an intellectual level what it is to have a handicap, enough to know that we should not want this embryo to develop into a person with such a handicap. But with future offspring the benefit seems particularly remote. Any possible benefit will accrue to future generations.

The problems concerning whether and to what extent we have obligations to future generations are well known. From a utilitarian point of view, however, we know that future people, whoever they are, will have preferences, and these cannot simply be disregarded. We have argued here that when what we are talking about is the production of new people, there are grounds for thinking that their preference-satisfaction should be a priority. But it might be argued that thinking in terms of the next generation is one thing, to give attention to more remote descendants is another. It is beyond the scope of this discussion to inquire into the question as to how far our obligations extend. But if we think it important to avoid pollution of the environment with nuclear waste, for the benefit of remote future generations, why should it be so difficult in the case of genes? In that it involves aiming at a stock of good genes in the gene pool, and in so far as it aspires to the elimination of the bad genes, eugenics seems to have something in common with the question of conservation of resources and avoidance of pollution, both of which are at the forefront of ecological discussions.

So let us turn to the risk of harm. Apart from the risk to the embryo, the risk to future generations consists in the possibility that the gene pool, which Anderson describes as the common heritage of all mankind, may be polluted by bad mistakes we make now.

It does not seem out of the question that we may think this risk worth taking in exchange for benefits to presently existing embryos and future generations in terms of genetic health.

However, Anderson prefers to make it more like the voluntary risk question and to ask the public to be aware and to sanction the experiments with germ line therapy.

· But the line of thought here seems to be not that we would be acting as proxies for future generations in volunteering to undergo the risk. It is rather that he sees the issue in terms of the risk to something of *ours*, viz. our heritage. In other words, *we* stand to lose something, while future generations will reap the benefits, if any.

It is not clear that this is the most useful way of looking at the problem. While many people care about the genes of their own children, it is not clear that so many care about the future of the species in an abstract sense. The really difficult questions would arise if the process of gene therapy involved risks of harm to presently existing people. When we think of the latter, however, it is not clear that somatic and germ line therapy are so different in the scope of their various risks. Anderson mentions the possibility of a viral vector, used in somatic therapy, being made pathogenic. This would expose the wider public to risks, and possibly their offspring too, as once a new organism has been created, it is there; how can we get rid of it again?

So perhaps it is the nature of the possible *benefit* that really makes the difference between somatic and germ line therapy. It may be that we do not care sufficiently about the embryos or that we can see no point in carrying out therapy which is going to affect their remoter offspring at a stage when they are not even a twinkle in their father's eye.

Robert Edwards has certainly taken the view that it would be better to destroy an embryo than to try to cure it. The practical problems are real and immediate, and the benefits remote and uncertain.

Anderson's solution to this is preliminary animal studies combined with a cost-benefit analysis approach.

But critics of genetic manipulation may say there must be an alternative way of dealing with the problem of genetic disease.

It might be suggested that genetic diseases should be treated phenotypically, as other diseases are. However, while we may alleviate the symptoms of genetic disease in the phenotype, we can not cure the genetic problem.

As we have seen, one suggested alternative for dealing with genetic diseases is to carry out screening programmes on foetuses, and abort those foetuses which have a genetic handicap. Screening has risks of its own: with amniocentesis, for example, there is the risk that the foetus, if healthy, may be born deformed by accident; abortion may be induced where it was not intended, or wrong diagnosis may result in an abortion where none was necessary.

However, the probability of these occurrences is statistically quite low and they apply only to foetuses.[49] All foetuses involved in the procedure suffer risk of death, and those which are found to be defective face certain death.

It may be true that the side effects of killing a foetus are less serious than those of killing an adult, but it is important to remember that abortion is something that stands in need of justification. The fact that one is carrying a genetically defective foetus would usually be regarded as one of the best reasons for having an abortion. But if that is the only reason for the abortion and there is the chance that the child could be given normal life as the result of gene therapy, it is far from clear that it would be right to have an abortion. It could be argued however that it is a policy which both avoids a risk to the gene pool and has the promise of high return, viz. the lack of sufferers from genetic disease.

Given that what we desire is the absence of suffering from genetic disease, however, it is not possible to say that it does not matter that we achieve this by killing the victims rather than by trying to cure them. To say this is to adopt R. N. Smart's *reductio ad absurdum* of negative utilitarianism that if what one is aiming at is the elimination of misery and suffering, one must be committed to killing everyone painlessly.[50] It needs to be shown that the undesirable consequences of the possible risks of gene therapy are worse than the killing of handicapped foetuses.

As far as embryos are concerned, suppose several embryos are produced in the course of *in vitro* fertilisation, and one is found to be handicapped. It seems inconceivable that a practitioner would choose to return that one to the woman. In making this decision the practitioner is making a eugenic judgement. It seems inevitable that this sort of judgement will be made increasingly as techniques improve. The consequences for this embryo will be either destruction or use in research. Would it be better, morally speaking, to try to cure all embryos? Those who think that the embryo should be entitled to the same respect as other human life would presumably say yes. From a consequentialist point of view it has to be admitted that the side effects of destroying an embryo are less serious than those of abortion, because for one thing the woman is not in a state of preparedness to bear, as she is in the case of abortion. The practice of multiple production of embryos also encourages the view that they are replaceable. The extent to which we think it urgent to try to overcome the practical difficulties will depend on the view we take of the embryo, but in order to overcome the practical difficulties, it may be that many embryos would have to be sacrificed in experiments anyway.

For some it may be in itself an argument against *in vitro* fertilisation that it makes eugenic decisions of this sort unavoidable.

The negative-positive distinction

We have looked at the preference not to have handicapped babies, and at different ways of bringing this about, along with the problems they raise. Some eugenists want to go further and advocate positive eugenics as well as negative. Thus Robert Sinsheimer talks of raising the human species to the highest genetic level, and C. K. Chan describes a scheme to encourage reproduction of those thought to be of above average intelligence in the population of Singapore. We have already mentioned, also, the Repository for Germinal Choice.

There is, first, a conceptual problem about the difference between negative and positive eugenics and genetic engineering. They are defined in terms of eliminating genetic handicaps or defects from the species and from individuals, and making improvements to the species and to individuals, respectively. But although up to now we have used the terms 'handicap' or 'defect' as if these notions were unproblematic, our criteria for what counts as a handicap, or defect, are at least partly evaluative, and change according to circumstances.[51] Also, some would say that we just do not have *any* criteria for what would count as an improvement.[52]

Anderson, however, is clear about the distinction: 'Replacing a faulty part is different from trying to add something new to a normally functioning system.'

Again, there is a problem about what counts as 'normal', but as I have discussed these questions elsewhere[53] we shall continue with the broad distinction that Anderson makes. Although the criteria may not be precise, we can understand in general terms the distinction between negative and positive.

Are there any arguments which distinguish them from a moral point of view?

One fairly common argument is that there is a minimum acceptable level below which no one should be pressed. Several reasons could be given for holding such a view. One is intuition. However, theoretical backing is needed for such an intuition. A utilitarian might appeal to the undesirable direct and side effects of pushing people below the level, or point to the principle of diminishing marginal utility, and suggest that in fact far more utility will be gained by helping those under the level than by improving the lot of those above.

For unless one believes in the acts-omissions distinction, the reasons one has for not pushing someone below the minimum would also lead one to hold that one should not allow someone who was already under it, to remain there (although the side effects of course might be worse in the former type of case).

In the question of the choice between negative and positive eugenics and genetic engineering, the problem is whether we can allow anyone to remain under a minimum acceptable level while pushing ahead with positive policies.

Two difficulties arise: can we interpret the minimum acceptable level in terms of genetic well-being, and if so, what result would it give us?

It is difficult enough to establish a minimum acceptable level when one is talking about economic distribution. As Rescher points out:

> as a utility economy rises from a condition of scarcity to one of abundance . . . the minimum is raised from bare survival to at-least-modestly-pleasant survival to a share in the good life. The minimum acceptable level is thus constantly changing.[54]

The same will apply to any minimum acceptable level for genetic well-being, but we might examine where it would be reasonable to establish it at the present time.

Firstly, to place the level just where negative policies end and positive policies begin would be too high. It is impracticable to suppose that we should ever reach a stage when no defects are present. New mutations, and inadequate methods of e.g. screening, would make this impossible. We cannot place a minimum acceptable level at an unreachable point.

The level will therefore be at some point within the negative domain, and it could be argued that those under this level should be helped before either other negative or any positive policies are put into action.

It seems unlikely that anyone would suggest the minimum acceptable level as being bare survival. For at the present time most people suffering from genetic defects can be kept alive for some time by phenotypic treatment in any case. Having the level at this stage would not even cover chromosomal abnormalities such as Down's syndrome.

A compromise could be sought by trying to find a level of preference-satisfaction at which we would want to place the level. We could approach this by seeking to divide the various defects into categories according to the probabilities they have of decreasing the sufferer's ability to achieve preference-satisfaction, and then giving priority to eliminating the worst ones from the species.

This seems an extremely difficult task. How could one compare the effect, for example, of a dominant disorder such as Huntington's chorea and a chromosome aberration such as Down's syndrome?

While difficult, it might be thought to be not theoretically impossible. However, it seems that even if we could establish that such a chromosome abnormality allows a lower probability of satisfaction than Huntington's chorea, to establish a minimum acceptable level between the two would be unsatisfactory. This is partly because it is necessary to take into consideration not simply the effect these defects have on their sufferers, but also the practicalities of eliminating them. Even if it were true that to suffer from Down's syndrome is far worse, it is nevertheless the case that this chromosome aberration is at present irremediable, and it is not possible to breed it out. The only way to eliminate it is by amniocentesis and abortion. It is not treatable by gene therapy.

Any attempt to establish a level must bear practicalities in mind. It would not be sensible to use up all resources trying to eliminate one group of defects while neglecting the possibility that other simple gene disorders, even if they are less serious, may be curable.

Secondly, the whole attempt to draw such lines between defects seems unrealistic. We do not follow this policy with diseases of the phenotype. Coughs and colds are treated along with typhoid.

Devoting all time to eliminating certain diseases might lead us into the trap of the law of diminishing returns. It might be counter-productive also to ignore the possibility of simple positive interventions.

Negative utilitarianism

This doctrine emphasises the minimisation of suffering rather than the maximisation of happiness.

It is not clear how it is to be interpreted (e.g. what counts as 'suffering'), but at first sight it might appear that the doctrine will tell us to give priority to removing defects which cause actual suffering, rather than eliminating other defects or making improvements.

However, it will only tell us always to do this if it gives no weight at all to classical utility. James Griffin argues that this is implausible and points out that if weight is given to classical utility, people who are not suffering may be helped in preference to those who are when the best results can be gained thereby, e.g. if there is a choice between making a large improvement for someone who is 'normal' and making a minimal difference to someone who has a defect.[55]

Equality

One argument for negative utilitarianism is that we should not widen the felicific gap.[56] Is it the case that the principle of equality would support a policy of giving priority to negative eugenics and genetic engineering?

Egalitarian principles come in different guises, but one thing at least is clear: they do not usually see genetic inequalities as undesirable *per se*. There seems to be widespread agreement that what we should be concerned with is not genetic equality, for two reasons: (a) that these inequalities are the result of nature; and (b) the principle of equality is a moral principle, and is not founded on biological fact. This moral principle may state no more than that everyone is owed equal respect, or it may be the utilitarian principle that everyone should count for one and no one for more than one. There may be a stronger attempt, however, to achieve equality in distribution of certain goods, or equality of quality of life.

Nothing follows from (a), however, to the effect that these

inequalities are desirable. They have been accepted because in the past nothing could be done about them. As regards (b) it is true that egalitarians have said that people are in some sense equal, or should be treated equally, despite their genetic differences, but does this show that it is not necessary to try to even out genetic differences?

One objection to such an idea is that we need biological variation in order that the species should have a good chance of survival. This is true, but the fact that we need variation does not show that it would be wrong to try to lessen genetic inequalities at all. We could still have a variety of talents and traits, along with different races and sexes.

Rawls says 'It is in the interest of each to have greater natural assets.'[57] What does he mean by this? He does not seem to mean that the individual desires greater assets, which would be false, or that if he had them he would be glad, which might also be false. What he means is presumably that in the conditions of choice supplied by the original position it would be rational to choose a better set of natural assets than a worse one (ignoring the fact that one set, although better overall, might be worse in some respect than the alternative), as being more likely to enable the possessor to obtain a share in the distribution of social goods.

Genetic resources can play an enormous part in the achievement of quality of life. The egalitarian who believes in equalising quality of life therefore has some reason to minimise genetic differences. In the past, when genetic means were not available, he might have thought the only way to achieve equal quality of life was by some form of compensation to those genetically disadvantaged, but if the genetic methods are available he has some reason to use them.

However, this egalitarian will not always take the view that negative eugenics and genetic engineering must take priority. Suppose that there are two people who have the same genetic make-up. Perhaps they are identical twins. But they have achieved vastly different qualities of life. In the science-fictional event that it be possible, would it be permissible to give the less well-off person a boost in his genetic capacities in order to give him a better chance in future?

If the egalitarian desires genetic equality only as an instrumental good, and not as an end in itself, the answer could presumably be yes. There may be circumstances, then, when he may support genetic inequality, and the achievement of this by a positive policy. The person of average height without growth hormone may be worse off as far as quality of life goes than the undersized midget who is quite happy at this height.

We have said that greater genetic resources make it more likely that preferences will be satisfied. The preference-utilitarian therefore has a reason to aim at higher genetic resources. He does not necessarily have a reason for equalising them, nor for giving priority to negative policies.

Moral solutions?

However, when we look at the various means available we can see problems about putting positive policies into practice. Thus some of the means will be open to the charge of elitism as described above. But to take the example of AID, if we should think it wrong for a practitioner to use sperm that he knew carried the gene for Huntington's chorea, why should it not be a good idea to use sperm carrying a predisposition for athletic ability, if that were desired?

The problem with AID, as we noted in Part I, is that it involves donation.

So we seem to be left with the positive version of gene therapy, or 'genetic enhancement'. W. French Anderson points out that this is already being done, giving the example of parents choosing to have growth hormone inserted into 'normal' children to make them specially good at basketball, etc. It seems to be this kind of thing (i.e. genetic enhancement) that Sinsheimer has in mind when he imagines that the new eugenics can raise all of mankind to the highest genetic level, while avoiding the social problems of the old eugenics.

If this were done at the somatic stage, presumably the same issues of consent would arise as with any other medical intervention.

The problems arise if we consider enhancement at the embryonic stage with subsequent consequences for the germ line.

Public policy

What are the issues of public policy that arise here?

The new eugenics does seem to avoid some of the objections to the old eugenics, in that it is open to parental choice, whereas the old eugenics was largely associated with government operated schemes. We have also seen that it can be for the benefit of the child to have a higher, rather than a lower, genetic potential.

However, the question arises as to what extent parental desires should be satisfied. Jonathan Glover has argued that a certain amount of censorship would have to be implemented to prevent people from choosing to handicap their children by e.g. selecting deafness.[58] This fits in with the approach taken here, which has been a child-centred one in which the child's chances of preference-satisfaction are given priority.

But in addition to operating selection at the level of the desires we also have to consider to whom it is to be open, and how widespread the practice should be, how it would work. Perhaps we can imagine an extension to genetic counselling services, so that couples could go for advice on how to have the child of their choice. But we must remember to be wary of producing a society that nobody wants by satisfying immediate preferences.

Just as some fear a society where children are seen as products, some fear a society in which genetic models predominate. We have looked at some of the thinking behind this. It is thought that when there is greater knowledge about genetics people will want to check their prospective partners for possible genetic problems, and this will have a detrimental effect on romance. But this has always been done in a rough and ready way. Family background has always had a tremendous effect on mating habits, even in societies where marriages are not officially arranged. (As long ago as the sixth century BC the Greek poet Theognis was lamenting that people were paying too little attention to questions of stock and too much to money.) More attention to genetic facts would only make it more professional.

Gene therapy and genetic enhancement would make it possible to by-pass the parental assessment to a certain extent in the knowledge that it might be possible to make adjustments to the children produced, if so desired.

Identity

However, we must remember that earlier we placed great importance on the question of genetic identity. In his article 'The idea of equality' Bernard Williams discussed the possibility of genetic manipulation to produce greater equality and said: 'here we might think that our notion of personal identity itself was beginning to give way; we might well wonder *who were* the people . . .'[59] If it became possible for parents to make choices about their children's genes when eggs had been fertilised, to what extent could we say that they had been provided with children *of their own*, and the children with parents of their own? It might be felt that if the genes are manipulated in the embryo, the children that will eventually develop will be different people from those who would have been produced if the manipulation had not taken place.

Thus having children 'of one's own' could become more a matter of degree than it is now. Instead of saying 'She's got your looks and my brains', parents may have to attribute these qualities to the expertise of the practitioner.

To what extent this would matter, to both parents and children, may depend on how drastic the alteration is, and on what particular question is being considered. It may be that they will be worried about a philosophical question of personal identity. If a new gene is added to X's genome, is the result still X or somebody else? Is the change different in kind from a kidney transplant?

However, these may not be the questions that trouble an individual who discovers that her parents had an extra gene or genes added. She may suffer from what today in the 'problem pages' is called an 'identity crisis', concerned with who she is, where she's coming from, and where she is going. Part of this may be an uncertainty about her genetic history. We have stressed the importance of this knowledge, and pointed out that when one does not know where 50 per cent of one's genes comes from, it can cause unhappiness. Would this problem be avoided if only a small amount of one's genetic make-up came from a source other than those who are rearing one?

This question is complicated by the fact that when one is ignorant as to who provided the sperm with its contribution to one's identity, the unhappiness may not be just a reaction to lack of knowledge of a certain proportion of one's genes but also caused by a sense of being abandoned by a person who has begotten, but not taken part in rearing.

Thus if that link has not been broken, if one knows who provided sperm and egg, if they are rearing, but one knows that an extra gene has been introduced into one's make-up to cure a genetic disease, there may be no cause for concern.

However, the greater the extent to which genetic changes are made, the less the child will be able to see herself as the child of her parents, and the more likely an identity crisis will be. This may depend not only on the quantitative aspect of the change but also on the qualitative aspect.

The essential question is whether any change we might make will be likely to increase the chances of quality of life for the people produced. If we can be confident of this in the case of curing defects, why not in the case of making improvements? But we must also remember that the more unnatural to our kind a change seems, the more cautious we must be in accepting that it is likely to have that result, and we must also be cautious about producing a situation where children either feel they do not really belong anywhere, because their genetic history is confused, or they feel alienated because they have been designed to suit their parents' aspirations, not their own. There are good reasons, as we have pointed out, for maintaining the link between begetting and rearing.

Sex choice and the sex ratio

Dharma Kumar's paper deals with one of the most controversial aspects of the new technologies, the issue of sex choice. 'Sex selection', as Warnock pointed out, is a term with two different meanings — the selection may be carried out either before or after fertilisation — either to establish the sex of an already existing embryo or to predetermine the sex of one yet to be produced.[60]

Sometimes it is advantageous to establish the sex of a foetus in order to assess the possibility of handicap, for example where there is a risk of transmitting a sex-linked hereditary disease such as haemophilia.

In vitro fertilisation has brought with it the possibility of further developments in sex selection. Testing could be carried out at a very early stage on embryos fertilised *in vitro*, and embryos only of the desired sex returned to the mother for development.

There is evidence to suggest that techniques such as amniocentesis have been used not solely to avoid the risk of sex-linked hereditary diseases, but also in a deliberate attempt to reduce the number of female children being born.[61] Dharma Kumar also draws our attention to the prevalence of female infanticide.

If there are good reasons for avoiding genetic handicap, then there may be reasons for using sex selection as a means to this end, depending on what the alternatives are.

But sex selection for reasons other than the avoidance of genetic handicap may have far-reaching social implications. Warnock says:

if an efficient and easy method of ensuring the conception of a child of a particular sex became available, it is likely that some couples would wish to make use of it for purely social reasons. Such a practice would obviously affect the individual family and the children involved, and would also have implications for society as a whole.[62]

Warnock does not spell out what these implications are likely to be. Is it possible to do so?

It is of course largely a matter of speculation as to what society would be like if the sex ratio became grossly unbalanced. We have science fiction accounts, but do we have any real guidelines to go on?

First of all, it needs to be made clear in which direction it is envisaged the choice will go. Most commentators seem to think that the overwhelming preference will be for boys. Thus Roberta Steinbacher and Helen B. Holmes suggest:

> However devalued, controlled, feared or exploited women have been, their indispensability to the continuation of the human race has remained a stubborn fact, conceded in even the most oppressive, patriarchal societies. However, now for the first time in human history, the power is at hand to negate that indispensability . . . the ultimate extension of the capacity to control embryonic development outside the female body is now being developed; artificial laboratory wombs . . . There is, to be blunt, the possibility of femicide.[63]

John Harris is very much in the minority in his discussion of the elimination of men and the production of the all-female world.[64]

What, if anything, would be wrong with the elimination of one sex? I shall consider the question, as Kumar does, largely with regard to the speculation that women would not be the popular sex.

Let us suppose, first, that the resulting world will have only one sex, not as in Ursula K. Le Guin's *The left hand of darkness* where individuals are not in normal circumstances distinguished sexually, but where as in our world they are either men or women.

The indispensability argument

Are either women or men indispensable? Traditionally of course they both were indispensable for the continuation of the species, but the advent of reproductive technology has brought with it the possibility that neither sex is any longer necessary for that purpose. Sperm banks, egg donation, and the possibility of artificial wombs have changed the character of human reproduction irrevocably.

Kumar takes a rather optimistic view of the indispensability of women, saying that women compare favourably in this respect with racial and religious minorities.

Holmes and Steinbacher, however, see women as no longer indispensable if men can reproduce without them. This depends on the view that they have no other essential contribution to make which cannot be duplicated.

This may be true if 'essential' is taken to be 'necessary for the continuance of the species'. But to grant this is not to admit that they have nothing else of value which they uniquely can offer. It may plausibly be thought that life would be impoverished if there were only one sex. A world in which 'masculine' values were not balanced by 'feminine' ones could be much worse than one in which both were in evidence. This of course depends on a view of what masculine and feminine values are.

Masculine and feminine values

Those who see power, competition, and control as essentially masculine fear the disappearance of the purported feminine values of co-operation and supportiveness. Thus pictures of what life would be like without women tend to be rather bleak.

But this argument assumes that what we think of as feminine values would disappear along with women. Is this so?

Joanna Russ describes a hypothetical Manland society in which there are no women:

There, in ascetic and healthful settlements in the country, little boys are made into Men — though some don't quite make it; sex change surgery begins at sixteen. One out of seven fails early and makes the whole change; one out of seven fails later and (refusing surgery) makes only half a change: artists, illusionists, impressionists of femininity who keep their genitalia but who grow slim, grow languid, grow emotional and feminine, all this the effect of spirit only. Five out of seven Manlanders make it; these are the 'real-men'.

129

The others are the 'changed' or 'the half-changed'. All real-men like the changed; some real-men like the half-changed; none of the real-men like real-men, for that would be abnormal. Nobody asks the changed or the half-changed what they like.[65]

If it is the case that gender differences are a social construct and do not reflect sex differences, then it is conceivable that such gender differences could survive the elimination of sex differences. Similarly in a manless society some women might become very masculine.

Thus it is assumed in Harris's picture of the setting up of the all-female world that the feminist project is to establish a world in which feminine values of co-operation and non-aggression will prevail. But as he points out, there is no guarantee that new forms of competition will not emerge.[66]

The self-defeating argument

The above argument suggests that sex choice might be self-defeating, but it might also be self-defeating in another way.

It is assumed that parents will make certain choices because they hope for particular results. Where people choose males they do so because they imagine that males are more advantaged or advantageous to their parents in that society. But this advantage will disappear if everyone is male.

However, it may not be a sufficient argument against free sex choice to argue that it will be self-defeating if everyone chooses the same. More argument would be needed to show that morally speaking, sex selection is undesirable.

The utilitarian argument

Kumar approaches the argument in a utilitarian way, asking about the effects on society and on the women concerned, and argues that as far as the effects on the girls produced are concerned, it might be better to be aborted than to be treated badly by their parents. As far as society is concerned, would a society with a more equally balanced sex ratio be a happier one?

This will depend partly on people's preferences. Men and

women are quite likely to have different preferences here. Their preferences will depend partly on what they see as the probable result of being in the majority or the minority. If women were in the majority would they be in harems or in positions of power? It is not inconceivable that men may prefer, if given the choice, to live in a society in which there were more women than men, where they could have their pick and live as a privileged minority, with harems and women servants to do the work. Women, however, may find the thought of being in the minority less attractive, if they think that it might lead to a situation where they were jealously guarded and restricted. Our experience suggests that these scenarios are not unlikely, though of course there are alternative ones. So people's preferences will at least partly depend on social and political arrangements.

Reflecting on this we can see that an equal ratio does not guarantee happiness. Equality in numbers will not satisfy the preferences of a sex which is oppressed in society. Thus women have been classed as a minority group not because of their number but because of their position in relation to the dominant group in society, i.e. men.

Thus Kumar takes the view that we need to change attitudes towards women in society, not prevent sex choice. There is still a question, however, about whether it would be easier to change those attitudes if the ratio were equal rather than weighted in one direction. Thus if we want to change attitudes towards women, we may need both to attempt to change them by means other than reproductive technology and to control sex choice.

However, even if we decide to keep the sex ratio equal, that is not the end of the story. Evidence suggests that the overwhelming preference, even when parents want a child of both sexes, is to have the boy first.[67] Given the purported correlation between one's position as firstborn and success in later life, such a use of sex choice could reinforce women's dependent position in society. Thus there may be reasons for adopting a policy of reverse discrimination here and trying to ensure that parents have girls first. Thus sex choice, far from being used as an expression of the low value placed on women in society, could be a means of redressing the balance, while still maintaining an overall equal ratio of numbers.

Of course, the problem is that it seems unlikely that this is a course which will recommend itself to men. If we follow the policy that we must take people's preferences as expressed, then we have to face the fact that both women and men at present express a preference for a boy child, especially as the firstborn. However, we also have to remember that we have preferences for living in certain sorts of society, and here we have to consider both equality in numbers and equality in the value placed upon men and women. Men and women may feel very differently about these, especially about the issue of reverse discrimination.

Are there any other reasons in favour of an equal ratio? To say that everyone would be given the best chance, in a society with representation of the sexes in equal numbers, of having a partner of the opposite sex, makes a lot of assumptions about the ways in which people want to relate. If 60 per cent of men were homosexual, without being bisexual, and 60 per cent of women were heterosexual, without being willing to share a partner, there could obviously be large numbers of women whose preferences were not being satisfied.

It would be nice to think that the best situation for both men and women would be a society in which they were roughly equal in numbers, and in which they were both equally valued. It is unfortunately difficult to be confident about this.

However, as we said when looking at the desire to have children, it would be unwise for society simply to say let everyone have their preference satisfied and then sort out the problems, if any, later. We have, again, to take a social policy decision on sex choice. The options are to refuse to allow sex choice except for genetic reasons, to allow choice subject to mechanisms designed to ensure equality, as in the voucher system described by Kumar, and deliberately weighting the choice in one direction, as in the reverse discrimination suggestion. Or, of course, we could refuse to allow sex choice at all.

Of these the second seems to allow both parental choice while maintaining social balance, and thus, it is to be hoped, the greatest chance of preference-satisfaction for members of both sexes. One could alter this suggestion slightly so that one couple could only choose a boy as firstborn if another chose a girl as firstborn, while avoiding the resentment caused by a more radical policy of reverse discrimination.

Conclusions

We have argued that the desire for a child of a certain quality must be exercised within certain limits, as must the desire for a child. The limits relate to the consequences for society and, in particular, for the child produced. What will be the likely effect of the child's genetic potential on her quality of life? Will she feel she belongs to no one because her genetic history is confused? And will she be required to live in a society where there are hardly any other women?

Notes and references

1. Plato, *Republic*, in *Platonis opera* (Oxford, Clarendon Press, 1957), vol. IV, 460a–c.

2. For the Arthur and Baby Doe cases and for the argument against the sanctity of life doctrine, cf. H. Kuhse and P. Singer, *Should the baby live?: the problem of handicapped infants* (Oxford University Press, Oxford, 1985).

3. J. K. Mason and R. A. McCall Smith, *Law and medical ethics* (Butterworth, London, 1983), p. 74.

4. *Buck* v. *Bell* 274 US 200 (1927), p. 207.

5. M. Warnock (Chairman), *Report of the Committee of Inquiry into Human Fertilisation and Embryology* (The Warnock Report), Cmnd. 9314 (HMSO, London, 1984), para. 11.22.

6. Ibid., para. 12.16.

7. Jeremy Rifkin, 'Perils of genetic engineering', *Resurgence*, vol. 109 (1985), pp. 4–7.

8. Errol E. Harris, 'Respect for persons' in R. T. de George (ed.), *Ethics and society: original essays on contemporary moral problems* (Macmillan, London, 1968), pp. 111–32.

9. A. Etzioni, *Genetic fix: the next technological revolution* (Harper Colophon, New York, 1975), p. 109.

10. Cf. Germaine Greer's account of the eugenics movement in *Sex and destiny* (Secker & Warburg, London, 1984), pp. 255–94.

11. *McKay* v.*Essex Area Health Authority* [1982] 2 All ER 771, [1982] 2 WLR 90, CA.

12. 'Nobel no-go, but still a quest for genius', *The Times*, 24 Sept. 1982.

13. M. W. Strickberger, *Genetics* (Macmillan, New York, 1968), Part 1.

14. J. D. Watson and F. H. C. Crick, 'A structure for deoxyribose nucleic acid', *Nature*, vol. 171 (1953), pp. 737–8.

15. A. E. Garrod, *Inborn errors of metabolism* (Oxford University Press, Oxford, 1909).

16. V. M. Ingram, 'Gene mutations in human haemoglobin: the chemical difference between normal and sickle cell haemoglobin', *Nature*, vol. 180 (1957), pp. 326–8.

17. C. Yanofsky *et al.*, 'On the colinearity of gene structure and protein structure', *Proceedings of the National Academy of Sciences*, vol. 51 (1964), pp. 266–72.

18. F. H. C. Crick *et al.*, 'General nature of the genetic code for proteins', *Nature*, vol. 192 (1961), pp. 1227–32.

19. J. Beckwith and J. King, 'The XYY syndrome: a dangerous myth', *New Scientist*, 14 Nov. 1974, pp. 474–6.

20. E.g. chorion villus sampling. Cf. 'Curing genetic diseases before birth', *New Scientist*, 12 April 1984.

21. Warnock, para. 9.3.

22. D. Bromhall, 'The great cloning hoax', *New Statesman*, 2 June 1978.

23. R. G. Edwards, 'The case for studying human embryos and their constituent tissues *in vitro*' in R. G. Edwards and J. M. Purdy (eds), *Human conception in vitro* (Academic Press, London, 1982), pp. 371–88.

24. 'Horizon: Brave new babies', BBC Television, 15 Nov. 1982.

25. *Buck* v. *Bell; Jacobson* v. *Massachusetts*, 197 US 11 (1905).

26. Etzioni, *Genetic fix*, p. 109.

27. T. Powledge, 'Can genetic screening prevent occupational disease?', *New Scientist*, 2 Sept. 1976, pp. 486–8.

28. This is partly because changes in the environment in themselves bring about changes at the genetic level, e.g. in influencing which genes are selected for, or in 'switching them on'. But here I am mainly concerned with moral differences.

29. M. Lappé, 'Reflections on the cost of doing science', *Annals of the New York Academy of Sciences*, vol. 265 (1976), pp. 102–11.

30. Philip Reilly, 'State supported mass genetic screening programs' in A. Milunsky and G. J. Annas (eds), *Genetics and the law* (Plenum Press, New York, 1976), pp. 159–84.

31. Beckwith and King, 'XYY syndrome'.

32. Lappé, 'Reflections'.

33. Ibid.

34. Ernest B. Hook, 'Geneticophobia and the implications of screening for the XYY genotype in newborn infants' in Milunsky and Annas, *Genetics and the law*, pp. 73–86.

35. Cf. Z. Harsanyi and R. Hutton, *Genetic prophecy: beyond the double helix* (Granada, London, 1982).

36. 'Lee puts hope in "Designer genes"', *Observer*, 18 Sept. 1983.

37. Hugh Lafollette, 'Licensing parents', *Philosophy and Public Affairs*, vol. 9, no. 2 (1980), pp. 182–97.

38. 'Take genes test, couples urged', *Guardian*, 28 Aug. 1985.

39. Unless one takes the view that there is no difference, in principle, between killing and preventing conception. Cf. J. Glover, *Causing death and saving lives* (Penguin, Harmondsworth, 1977), pp. 60–73.

40. See Kuhse and Singer, *Should the baby live?*, pp. 98–117, for evidence to suggest that infants can also be kept in a separate category, psychologically.

41. G. J. Annas, 'The case of Elizabeth Bouvia', *Hastings Center Report*, vol. 14, no. 2 (1984), pp. 20–3.

42. C. Rivers, 'Genetic engineering portends a grave new world' in Thomas R. Mertens (ed.), *Human genetics: readings on the implications of genetic engineering* (John Wiley, New York, 1975), pp. 7–15.

43. Plato, *Republic*, 432b–434e.

44. R. G. Edwards, 'Fertilization of human eggs *in vitro*: morals, ethics and the law', *Quarterly Review of Biology*, vol. 49 (1974), pp. 3–26.

45. J. Passmore, *Man's responsibility for nature* (Duckworth, London, 1974), p. 239.

46. E.g. A. R. Manser, 'The concept of evolution', *Philosophy*, vol. 40 (1965), pp. 18–34.

47. A. G. Motulsky, 'Impact of genetic manipulation on society and medicine', *Science*, vol. 219 (1983), pp. 135–40.

48. Lord Rothschild, 'Risk', *The Listener*, 30 Nov. 1978, p. 717.

49. Set at 'between 1 and 5 per cent, depending on the surgeon' in R. M. Restak, *Premeditated man: bioethics and the control of future human life* (Penguin, New York, 1977), p. 76.

50. R. N. Smart, 'Negative utilitarianism', *Mind*, vol. 67. (1958), pp. 542–3.

51. Cf. D. Thomas, *The experience of handicap* (Methuen, London, 1982), pp. 3–20.

52. But cf. C. O. Carter, *Human heredity* (Penguin, Harmondsworth, 1962), p. 246, for a list of desirable qualities.

53. In 'Genetic improvement', forthcoming.

54. N. Rescher, *Distributive justice: a constructive critique of the utilitarian theory of distribution* (Bobbs-Merrill, New York, 1966), pp. 28–30.

55. J. Griffin, 'Is unhappiness morally more important than happiness?', *Philosophical Quarterly*, vol. 29 (1979), pp. 47–55.

56. A. D. M. Walker, 'Negative utilitarianism', *Mind*, vol. 83 (1974), pp. 424–8.

57. J. Rawls, *A theory of justice* (Oxford University Press, Oxford, 1973), p. 108.

58. J. Glover, *What sort of people should there be?* (Penguin, Harmondsworth, 1984), pp. 47–51.

59. B. Williams, 'The idea of equality' in *Problems of the self* (Cambridge University Press, Cambridge, 1973), p. 246.

60. Warnock, para. 9.4.

61. M. Kishwar, 'The continuing deficit of women in India and the impact of amniocentesis' in G. Corea *et al.*, *Man made women: how new reproductive technologies affect women* (Hutchinson, London, 1985), pp. 30–7.

62. Warnock Report, para. 9.11.

63. R. Steinbacher and H. B. Holmes, 'Sex choice: survival and sisterhood' in Corea *et al.*, *Man made women*, pp. 52–63.

64. J. Harris, *The value of life* (Routledge & Kegan Paul, London, 1985), pp. 166–73.

65. J. Russ, *The female man* (Women's Press, London, 1985), p. 167.

66. Harris, *Value of life*, p. 172.

67. Steinbacher and Holmes, 'Sex choice', p. 53.

7

The Prospect of Designed
Genetic Change

Robert L. Sinsheimer

It has now become a serious necessity to better the breed of the human race. The average citizen is too base for the everyday work of modern civilisation. Civilised man has become possessed of vaster powers than in old times for good or ill but has made no corresponding advance in wits and goodness to enable him to conduct his conduct rightly.

This was written in 1894 by Sir Francis Galton. The concerns of the present are clearly not new.

It has long been apparent that you and I do not enter this world as unformed clay compliant to any mould; rather, we have in our beginnings some bent of mind, some shade of character. The origin of this structure — of the fibre of this clay — was for centuries mysterious. In earlier times men sought its trace in the conjunction of the stars or perhaps in the momentary combination of the elements at nativity. Today, instead, we know to look within. We seek not in the stars but in our genes for the herald of our fate.

Today there is much talk about the possibility of human genetic modification — of designed genetic change, specifically of mankind. A new eugenics has arisen, based upon the dramatic increase in our understanding of the biochemistry of heredity and our comprehension of the craft and means of evolution. I think this possibility, which we now glimpse only in fragmented outline, is potentially one of the most important concepts to arise in the history of mankind. I can think of none with greater long-range implications for the future of our species. Indeed this concept marks a turning-point in the whole evolution of life. For the first time in all time a living creature understands its origin and can undertake to design its future. Even in the ancient myth man was constrained by his essence. He could not rise above his nature to chart his destiny. Today we can envision that chance — and its dark companion of awesome choice and responsibility.

It is all too easy, albeit useful, to let our imagination in these matters roam far beyond our technical base. It is easy, even for

136

modest men given to cautious projections, because in truth all that seems needed is the technology and the resolution to transfer to man what we already know to be feasible in bacteria or carrot cells or frogs. It is easy because there are no known natural laws to repeal or contravene. None of the time warps or hyper-drives or teleportation of science fiction are needed to envision vegetative reproduction, organ generation, genetic therapy, or eugenic transformation of our species.

I would like, however, to consider a very specific and possible use of our newer knowledge, relating to a major biomedical problem. This application may well seem of small dimensions as compared to some sweeping prospects, but I believe it will illuminate the state of our knowledge and our technology and will thereby reveal the shape of things to come.

I want to use the phrase 'genetic change' in a broad sense, in the sense of altering some physiological or psychological process which at present we believe has been programmed into us through our inheritance. And I will assume that such change might be achieved either in a strictly genetic mode through a change in our inherited characteristics, or in a somatic (non-inheritable) mode — possibly through a change in the time or place or degree of action of our inherited genetic components, or possibly through the somatic addition of genetic components. Obviously changes of the former — the truly genetic type — have the greater ultimate potential; for the very nature of the species seems potentially susceptible to change. Change of the latter type — somatic genetic modifications — are more limited. Their scope and function are the more restricted, but they are also undoubtedly the more accessible possibilities which we will first achieve.

There are in the United States today some 4,000,000 clinical diabetics. Many of these people are kept alive only by repeated, frequent injections of the hormone insulin. It is believed that there are several million more cases with marginal symptoms. Without recurrent injections of insulin many of these people would perish. While it keeps them alive, the injection of insulin is not the full equivalent of a normal physiological function; diabetics are known to be more susceptible to disease, to heart and circulatory illnesses, and other physical limitations than non-diabetics.

I propose that genetic therapy offers the promise of a much more elegant, and indeed more satisfactory, physiological solution to this ailment. And there are various possible genetic approaches.

To begin with we must understand the normal process of insulin

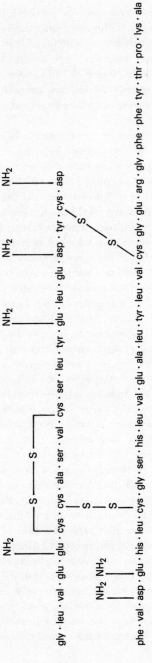

Figure 7.1: The chemical structure of human insulin consists of two polypeptide chains, one of 21 amino acids and the other of 30, which are joined by two disulphide bonds.

formation. Insulin is a protein, composed of two polypeptide chains — one of 21 amino acids and one of 30 — joined by two disulphide bonds. There is recent evidence that indicates strongly that the insulin molecule is initially formed as a single polypeptide chain, and an internal segment is subsequently excised by the action of a specific proteolytic enzyme (Figure 7.1).

The synthesis of this protein, the proinsulin, is accomplished in the usual manner: The hereditary instructions specifying the sequence of amino acids for insulin are encoded in a segment of the DNA from the cell nucleus (Figure 7.2.). The instructions are copied and transcribed into a messenger RNA molecule which then is transported out of the nucleus to the cytoplasm (Figure 7.3.). There protein synthesis takes place upon the ribosomes. In this process the sequence of nucleotides in the RNA is translated into the corresponding amino acid sequence with the help of the transfer RNA molecules, the activating enzymes, the initiators, the coupling factors, and all the rest of a very complex machinery.

Figure 7.2: This schematic drawing of the conversion of proinsulin to insulin illustrates the recent evidence that the insulin molecule is formed as a single polypeptide chain and that an internal segment is subsequently excised by the action of a specific proteolytic enzyme.

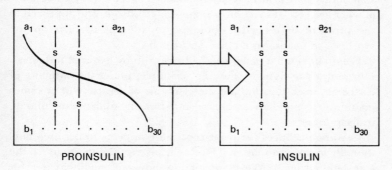

PROINSULIN   INSULIN

It is well known that this synthesis of insulin normally takes place only in the beta cells in the Islands of Langerhans in the human pancreas. In the diabetic these cells fail to produce an adequate amount of insulin. Now it is believed, and there is good reason for this belief from studies of lower animals, that the *full* DNA content of the genome is present in every somatic cell. And thus we believe that the genetic instructions specifying the sequence of proinsulin are present in all the cells of the body and not only the beta cells of the Islands of Langerhans. Evidently they are not activated, or, as it is more fashionable to assume these days, they are repressed. Repression could take place at any of several levels.

A typical somatic cell is only called upon to use a small fraction of its genome. There is good evidence that in a liver or a muscle cell no more than 5 per cent of the DNA is ever transcribed into RNA, so there is repression at the chromosomal level. Further, it is clear that perhaps half or more of that which *is* transcribed never reaches the cytoplasm to be translated. And even if the RNA reaches the cytoplasm, there is evidence for specific blocks at the translational level. There are clearly many opportunities for the restriction of expression of the inherited genetic instructions.

Figure 7.3: Steps in the biosynthesis of insulin. Repression of insulin synthesis could take place at any of these stages.

DNA → RNA(nucleus) → RNA(cytoplasm) → PROINSULIN → INSULIN
  Transcription   Transport   Translation   Activation

In the case of insulin we do not know by what means the expression of this gene is limited to a few islands of cells. We do not know at what level the restriction is imposed. However, one approach to the problem of diabetes would be to attempt to turn on the synthesis of insulin in another set of cells.

We do know that genes can be turned on by external influence. Hormones do this every day. For example, under the influence of cortisone, liver cells initiate the synthesis of a variety of enzymes including tryptophan pyrrolase and tyrosine-alpha-ketoglutarate transaminase.

In some instances the prior repression appears to be lifted hormonally at the chromosomal level of transcription, in others, at the translational level. We do not know how we might do this for insulin, but we can see a clear model. And in fact just such an activation or derepression for insulin *must* have occurred through some chain of ontogenetic events during embryonic development to activate — to turn on — the appropriate genes in the beta cells of the Islands of Langerhans.

We should not oversimplify this problem. Obviously if we want a new group of cells to synthesise insulin, we must not only activate the gene for proinsulin but also arrange for its conversion to insulin and for its release from the cells. But, if we were fortunate, these functions might all come as a genetic package.

There is a radically different genetic approach that we might take alternatively. Instead of an attempt to lift this profound repression of the expression of the gene for insulin, we might, in principle, supply to a group of cells a wholly new gene or set of genes which would code for the synthesis of insulin and which might not be subject to the normal somatic pattern of repression.

How might we add, in such a specific manner, to the genetic components of a cell? Our models come from studies with bacterial cells. In these organisms a variety of means exist to permit exchange and in so doing to provide small increments of genetic material. These include transformation, contact transfer both chromosomal and episomal, and transduction both general and specific. Organelles for contact transfer are not known among mammalian cells, and transformation as such has not yet been convincingly demonstrated in mammalian cells. Therefore, the possible use of transduction as a means to genetic modification of cells of higher organisms should be specifically considered.

Transduction among bacteria involves the transfer of genetic material, DNA, from one cell to another through viral mediation. I would like to present two particular cases.

The first case is that of the bacterial virus P1, which contains one molecule of DNA of about 60,000,000 in molecular weight. Upon infection of the cell by certain types of P1, the cell is lysed (broken down) after half an hour to produce a few hundred progeny virus particles. Most of these will contain a DNA identical to that of the virus that initiated the infection. However, a little less than 1 per cent of the particles will contain *instead* a piece of the DNA of the chromosome of the host bacterium, a piece also about 60,000,000 in molecular weight. Which piece of DNA — which particular 60,000,000 out of the 3 billion molecular weight of host DNA — is random. The particular virus will contain the piece carrying, say, genes D and E, while another carries a piece with the genes P and Q, etc.

By appropriate means these particles carrying host DNA, called transducing particles, can be separated from those carrying the normal viral DNA. When such transducing particles are added to susceptible bacteria, the DNA inside the virus particle is, in the usual way of bacterial viruses, injected into the cell. But now we have added to the cell not a destructive virus genome, but a piece of bacterial DNA which may well carry genetic markers not present in this particular host. This DNA may be transcribed at once to yield new protein.

For this piece of DNA to perpetuate itself, however, it must, in general, become incorporated *into* the host chromosome by a process of genetic recombination. Normal bacterial cells have the enzymatic machinery to do this, and, in the case of the transducing particles of phage P1, there is about one chance in ten that the particular piece of DNA will be so incorporated and perpetuated.

In bacterial cells there are often small secondary chromosomes — episomes — usually containing 1 or 2 per cent as much DNA as the principal chromosome. These are physically separate from the principal chromosome, but usually replicate in synchrony with it. It is possible for a P1 phage to pick up and transfer an entire episome as well as a piece of bacterial chromosome.

A second case of transduction concerns the temperate (frequently non-lethal) bacteriophage lambda. Upon infection with the bacteriophage lambda, the result in an appreciable percentage of the cells (it can be the majority) is the physical incorporation of either the viral DNA or of one of its descendants into the chromosomes of the host. Following this, the *virus-like* tendencies of this DNA are suppressed. The cell survives and multiplies, and the incorporated viral DNA is replicated into each daughter cell along with the rest of the bacterial chromosome. Such a virus-carrying cell is said to be a lysogen.

An important feature is that the point of insertion for the lambda DNA into the host DNA is specific, and it is determined by the particular virus which in turn specifies an enzyme — an integrase — which brings about its incorporation at that site. Related strains of lambda-type viruses are known which integrate into *other* chromosomal sites because they have different integrases.

It is possible, however, to induce an activation of this carried viral DNA in the lysogenic cells — to cause it to remember that it really is a virus, to cause it to break out of the bacterial chromosome, to begin to multiply, to produce progeny, and to lyse the cell, producing new virus particles.

Occasionally in such an activation, which is called induction, the piece of DNA which splits out of the chromosome is not strictly the viral genome but may incorporate a piece of the neighbouring bacterial chromosome with its genetic material in lieu of a piece of the viral genome. Under certain circumstances viral development can proceed anyway. Such pieces of DNA, partially viral and partially host, can multiply and can be incorporated into virus particles (Figure 7.4.). Particles with this mixed DNA can be isolated from the bulk of progeny. If they are now added to susceptible cells, this DNA can still integrate into the host chromosome at the same locus but now adding along with the viral genes a specific piece of DNA from the former host — which may carry specific novel genetic traits into the new host.

Figure 7.4: In excision of a lambda DNA from the host chromosome, normal excision (centre) yields one complete viral genome, while abnormal excision (left and right) yields mixed genomes, part viral and part host.

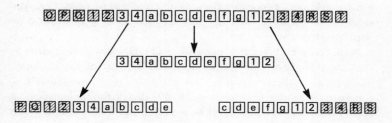

In both of these cases, then — P1 and lambda — the net result is the introduction, via a particle normally indistinguishable from a virus, of new genetic material into the host cell. In the first instance the new factors added are random relative to the host genome. In the second, they are factors found at specific sites near the normal region of integration into the virus. The region varies in different viruses.

Could a similar transfer be accomplished with a virus in the cells of higher organisms? We have every reason to think that it does occur. Upon infection of mammalian cells with the simian virus 40, or with polyoma virus, or with some strains of adenovirus, in a fraction of such infected cells the viral DNA becomes established within the cell. Whether it is integrated into the chromosomal DNA or is an episome is not known. It is then perpetuated within the clone of cells descended from the original infected cell, as in a lysogenic bacterium. The information carried in the viral DNA is certainly expressed: cells carrying such DNA have altered properties; that messenger RNA derives from this viral DNA can be demonstrated; new protein antigens have been detected within such cells; and in special circumstances the entire genome of the virus can be recovered (and hence must have been present) from remote descendants of such altered cells.

Technically and literally the stage is set. If we could obtain a virus analogous to simian virus 40 — able to persist within altered cells and carrying an expressible gene for proinsulin in lieu of a normal viral gene — we might indeed be able to provide a genetic alternative to the daily injection of insulin.

The problem then is, where are we to find this virus for propitiously carrying a gene to provide insulin? Such a virus might exist in nature, but I propose that we should quite literally, in time, be able to make it to order. We will have the ability in the not distant future to synthesise a polynucleotide chain capable of coding for insulin and for the other genes necessary to integrate the DNA into a chromosome, or to maintain it as an episome, or whatever. And we will then also be able to package this *de novo* DNA into an appropriate virus coat.

Is this pure fantasy? No, not really. The DNA of simian virus 40 consists of a chain of 5,000 nucleotides. The art of specific polynucleotide synthesis is young but thriving. It is now feasible to construct a specific sequence of 50 deoxyribonucleotides. A sequence of up to 100 seems close at hand, and a thousand or a few thousand is by no means inconceivable.

Furthermore, such a synthesis needs to be done only once. Once the DNA is available, nature provides the means to copy it with the highest fidelity.

Similarly, our understanding of the process of viral self-assembly is growing swiftly, and but a small step behind is the art of viral assembly *in vitro*. The technology needed for such a radically different approach to a major clinical problem is almost in reach.

Though the analogy is not perfect, in describing these prospects I feel strangely akin to the physicists who pointed out in the 1930s that the principles required for the release of the energy locked in the atomic nucleus were understood. All that was needed was a practical breakthrough and the requisite technology. Here, too, the principles seem in hand. All that really seems needed is optimism, sustained effort, and support commensurate with the importance of the problem.

The larger and the deeper challenges — those concerned with the defined genetic improvement of man — perhaps fortunately are not yet in our grasp, but they are etched clear upon the horizon. We should begin to prepare now for their reality.

It is worthwhile to consider specifically wherein the potential of the new genetics exceeds that of the old. To implement the older eugenics of Galton and his successors would have required a massive social programme carried out over many generations.

Such a programme could not have been initiated without the consent and co-operation of a major fraction of the population, and would have been continuously subject to social control. In contrast, the new eugenics could, at least in principle, be implemented on a quite individual basis, in one generation, and subject to no existing social restrictions.

The old eugenics would have required a continual selection for breeding of the fit, and a culling of the unfit. The new eugenics would permit in principle the conversion of all the unfit to the highest genetic level.

The old eugenics was limited to a numerical enhancement of the best of our existing gene pool. The horizons of the new eugenics are in principle boundless — for we should have the potential to create new genes and new qualities yet undreamed. But of course the ethical dilemma remains. What are the best qualities, and who shall choose?

It is a new horizon in the history of man. Some may smile and may feel that this is but a new version of the old dream, of the perfection of man. It is that, but it is something more. The old dreams of the cultural perfection of man were always sharply constrained by his inherent, inherited imperfections and limitations. Man is all too clearly an imperfect and flawed creature. Considering his evolution, it is hardly likely that he could be otherwise. To foster his better traits and to curb his worse by cultural means alone has always been, while clearly not impossible, in many instances most difficult. It has been an Archimedean attempt to move the world, but with the short arm of a lever. We now glimpse another route — the chance to ease the internal strains and heal the internal flaws directly, to carry on and consciously perfect far beyond our present vision this remarkable product of two billion years of evolution.

I know there are those who find this concept and the prospect repugnant — who fear, with reason, that we may unleash forces beyond human scale and who recoil from this responsibility. I would suggest to them that they do not see our present situation whole. They are not among the losers in that chromosomal lottery that so firmly channels our human destinies. This response does not come from the 250,000 children born each year in the USA with structural or functional defects, of which an estimated 80 per cent involve a genetic component. And this figure counts only those with gross evident defects outside those ranges we choose to call natural. It does not include the 50,000,000 'normal' Americans with an IQ of less than 90.

We are among those who were favoured in the chromosomal lottery, and, in the nature of things, it will be our very conscious choice whether as a species we will continue to accept the innumerable individual tragedies inherent in the outcome of this mindless, age-old throw of dice, or instead will shoulder the responsibility for intelligent genetic intervention.

As we enlarge man's freedom, we diminish his constraints and that which he must accept as given. Equality of opportunity is a noble aim given the currently inescapable genetic diversity of man. But what does equality of opportunity mean to the child born with an IQ of 50?

The application of knowledge requires technology, but the impact of knowledge can precede its application. Knowledge brings understanding, and the consequences of understanding can overflow the mind into the heart. It may be that in the near future the most important consequences of our new knowledge of ourselves will be a new sense of the power and responsibility — of the pivotal role — of man in this universe. Copernicus and Darwin demoted man from his bright glory at the focal point of the universe to be merely the current head of the animal line on an insignificant planet. In the mirror of our newer knowledge we can begin to see that in truth we are far more than another ephemeral form in the chain of evolution. Rather we are an historic innovation. We can be the agent of transition to a whole new pitch of evolution. This is a cosmic event.

8

Human Gene Therapy: Scientific and Ethical Considerations

W. French Anderson

There are four potential levels of the application of genetic engineering for the insertion of a gene into a human being:

(1) Somatic cell gene therapy: this would result in correcting a genetic defect in the somatic (i.e. body) cells of a patient.

(2) Germ line therapy: this would require the insertion of the gene into the reproductive tissue of the patient in such a way that the disorder in his or her offspring would also be corrected.

(3) Enhancement genetic engineering: this would involve the insertion of a gene to try to 'enhance' a known characteristic; for example, the placing of an additional growth hormone gene into a normal child.

(4) Eugenic genetic engineering: this is defined as the attempt to alter or 'improve' complex human traits each of which is coded by a large number of genes; for example, personality, intelligence, character, formation of body organs, and so on.

Somatic cell gene therapy

There are many examples of genes which, when defective, produce serious or lethal disease in a patient. Gene therapy should be beneficial primarily for the replacement of a defective or missing enzyme or protein that must function inside the cell that makes it, or of a deficient circulating protein whose level does not need to be exactly regularised (for example, blood clotting factor VIII which is deficient in haemophilia). Early attempts at gene therapy will almost certainly be done with genes for enzymes that have a simple

147

'always-on' type of regulation (For a technical discussion of the state-of-the-art of somatic cell gene therapy, together with extensive references, see Anderson, 1984).[1]

Initial candidates for gene therapy

The most likely genes to be used in the first experiments on human gene therapy are: hypoxanthine-guanine phosphoribosyl transferase (HPRT), the absence of which results in Lesch-Nyhan disease (a severe neurological disorder that includes uncontrollable self-mutilation); adenosine deaminase (ADA), the absence of which results in severe combined immunodeficiency disease (in which children have a greatly weakened resistance to infection and cannot survive the usual childhood diseases); and purine nucleoside phosphorylase (PNP), the absence of which results in another form of severe immunodeficiency disease. For all three, the clinical syndrome is profoundly debilitating. The disorder in each is found in the patient's bone marrow (although the severe central nervous system manifestations of Lesch-Nyhan disease are due to the absence of HPRT in brain cells and probably cannot be corrected with current techniques). In all three there is no, or minimal, detectable enzyme in marrow cells from patients who have no copies of the normal gene. In these patients the production of a small percentage of the normal enzyme level should be beneficial and a mild overproduction of enzyme should not be harmful. In addition, in the case of all three disorders the normal gene has been cloned and is available.

Previously, clinical investigators thought that the human genetic diseases most likely to be the initial ones successfully treated by gene therapy would be the haemoglobin abnormalities (specifically, beta-thalassaemia) because these disorders are the most obvious ones carried by blood cells, and bone marrow is the easiest tissue to manipulate outside the body. Regulation of globin synthesis, however, is unusually complicated. Not only are the embryonic, foetal, and adult globin chains carefully regulated during development, but also the subunits of the haemoglobin molecule are coded by genes on two different chromosomes. To understand the regulatory signals that control such a complicated system and to develop means for obtaining controlled expression of an exogenous (i.e. inserted by gene therapy) beta-globin gene will take considerably more research effort.

Severe combined immunodeficiency due to a defect in the ADA gene can be corrected by infusion of normal bone marrow cells from a histocompatible donor. Therefore, selective replication of the normal marrow cells appears to take place. This observation offers hope that defective bone marrow can be removed from a patient, the normal ADA gene inserted into a number of cells through gene therapy, and the treated marrow reimplanted into the patient where it may have a selective growth advantage. There is also evidence that marrow cells containing the normal gene for HPRT may have a selective advantage (in both mice and humans) over cells that do not. If selective growth occurs, elimination of the patient's own marrow would not be necessary. If, however, corrected marrow cells have no growth advantage over endogenous (i.e. the patient's own untreated) cells, then partial or complete marrow destruction (either by irradiation or by other means) may be required in order to allow the corrected marrow cells an environment favourable for expansion. The latter situation would require much greater confidence that the gene therapy procedure will work before a clinical trial should be undertaken.

Ethics

The ethics of gene therapy in humans has been discussed for many years and is being widely debated at present. John Fletcher has reviewed this area in detail (Fletcher, 1985).[2] Essentially all observers have stated that they believe that it would be ethical to insert genetic material into a human being for the sole purpose of medically correcting a severe genetic defect in that patient, in other words, somatic cell gene therapy. Attempts to correct a patient's reproductive cells (i.e. germ line gene therapy) or to alter or improve a 'normal' person by gene manipulation (i.e. enhancement or eugenic genetic engineering) are controversial areas. However, somatic cell gene therapy for a patient suffering a serious genetic disorder would be ethically acceptable if carried out under the same strict criteria that cover other new and experimental medical procedures. The techniques that are now being developed by clinical investigators for human application are for somatic cell, not germ line, gene therapy.

What criteria should be satisfied prior to the time that somatic cell gene therapy is tested in a clinical trial? Three general requirements, first presented in 1980 (Anderson and Fletcher, 1980),[3] are that it should be shown in animal studies that (i) the new gene can be put into the correct target cells and will remain there long enough to be effective; (ii) the new gene will be expressed in the cells at an appropriate level; and (iii) the new gene will not harm the cell or, by extension, the animal. These three requisites, summarised as delivery, expression and safety, will each be examined in turn.

These criteria are very similar to those required prior to the use of any new drug, therapeutic procedure, or surgical operation. The requirements simply state that the new treatment should get to the area of disease, correct it, and do more good than harm. Some flexibility is necessary since the criteria might be altered for a critically ill patient for whom no further conventional therapy is available. The exact definitions of what is 'long enough to be effective', what is an 'appropriate level', and how much harm is meant by 'harm', are questions for ongoing discussion as more is learned about gene therapy. Ultimately, local Institutional Review Boards and the National Institutes of Health (NIH), the latter through its newly created Working Group on Human Gene Therapy, must decide if a given protocol is ready for human application. Once the criteria are satisfied, that is, when the probable benefits for the patient are expected to exceed the possible risks, then attempts to cure human genetic disease by treatment with somatic cell gene therapy would be ethical. The goal of biomedical research is, and has always been, to alleviate human suffering. Gene therapy is a proper and logical part of that effort.

Delivery

At present, the only human tissue that can be used effectively for gene transfer is bone marrow. No other cells (except, perhaps, skin cells) can be extracted from the body, grown in culture to allow insertion of exogenous genes, and then successfully reimplanted into the patient from whom the tissue was taken. In the future, as more is learned regarding how to package the DNA and to make it tissue-specific, the intravenous route would be the simplest and most desirable. However, attempting to give a foreign gene by injection directly into the bloodstream is not advisable with our present state of knowledge since the procedure would be enormously inefficient and there would be little control over the DNA's fate.

Studies are considerably more advanced with bone marrow than skin cells as a recipient tissue for gene transfer. Bone marrow consists of a heterogeneous population of cells, most of which are committed to differentiate into red blood cells, white blood cells, platelets, and so on. Only a small proportion (0.1 to 0.5 per cent) of nucleated bone marrow cells are stem cells (that is, blood-forming cells that have not yet differentiated into specific cell types and which divide as needed to maintain the marrow population). In gene therapy, it would be these rare, unrecognisable stem cells that would be the primary target. Consequently a delivery system useful for gene therapy must be efficient.

Several techniques for transferring cloned genes into cells have been developed (Anderson, 1984).[4] Each procedure is valuable for certain types of experiments, but none can yet be used to insert a gene into a specific chromosomal site in a target cell. At present the most promising approach for use in humans employs retro-virus-based vectors carrying exogenous genes.

Vectors derived from retroviruses possess several advantages as a gene delivery system. First, up to 100 per cent of cells can be infected and can express the integrated viral (and exogenous) genes. Second, as many cells as desired can be infected simultaneously; 10^6 to 10^7 is a convenient number for a simple protocol. Third, under appropriate conditions the DNA can integrate as a single copy at a single, albeit random, site. Finally, the infection and long-term harbouring of a retroviral vector usually does not harm cells. Several retroviral vector systems have been developed; those projected for human use at the present time are constructed from the Moloney murine (mouse) leukaemia virus. Evidence obtained from studies with experimental animals and in tissue culture indicates that retroviruses can be used as a reasonably efficient delivery system.

An ideal delivery system would be tissue-specific. When a genetic disorder is in the blood cells, then the isolated bone marrow can be treated. But no other tissue (except skin cells) can be removed, treated, and replaced at present. Since many viruses are known to infect only specific tissues (that is, to bind to receptors that are present only on certain cell types), a retroviral particle containing a coat that recognises only human blood-forming cells would permit the retroviral vector to be given intra-venously with little danger that cells other than those in the marrow would be infected. In the future, such specificity could permit the liver and brain, for example, to be treated individually. In addition, the danger of inadvertently affecting germ cells could be eliminated. One problem, however, is that cell replication appears to be necessary for retrovirus integration. It would not be possible to infect non-dividing brain cells, for example, so far as we now know.

The optimal system not only would deliver the vector specifically into the cell type of choice but would also direct the vector to a predetermined chromosomal site. Specific insertion into a selected site on a chromosome can be achieved in lower organisms but has not yet been possible in mammals.

Expression

In order for gene therapy to be successful there must be appropriate expression of the new gene in the target cells. Even when a delivery system can transport an exogenous gene into the DNA of the correct cells of an organism, it has been a major problem to get the integrated DNA to function. A vast array of cloned genes have been introduced into a wide range of cells by several gene transfer techniques. 'Normal' expression of exogenous genes is the exception rather than the rule.

Expression of exogenous genes carried by retroviral vectors into intact animals via treated bone marrow cells has been reported by three laboratories. Two studies demonstrated the expression of an antibiotic resistance gene in mice (Joyner *et al.*, 1983; Williams *et al.*, 1984).[5] The most extensive data, however, are from studies with the enzyme HPRT (Miller *et al.*, 1984; Willis *et al.*, 1984).[6] A homozygous Lesch-Nyhan (LN) lymphoblast (white blood cell) line, which lacks a functional HPRT gene, was used to demonstrate that an HPRT human blood-forming cell could be corrected by a retroviral vector containing an active HPRT gene. In a corollary study, viral particles containing the HPRT-vector were used to infect mouse bone marrow cells that were then injected into lethally irradiated mice. Both human HPRT protein and chronic production of HPRT-vector particles were detected in the blood-forming tissues of the mice. These data provide hope that vectors can eventually be built with all the regulatory signals necessary to produce correctly controlled expression of exogenous genes in target cells.

Safety

Finally, a human gene therapy protocol must be safe. Although retroviruses have many advantages for gene transfer, they also have disadvantages. One problem is that they can rearrange their own structure as well as exchange sequences with other retroviruses. In the future it might be possible to modify non-infectious retroviral vectors in such a way that they remain stable. At present, however, there is the possibility that a retroviral vector might recombine with an endogenous viral sequence to produce an infectious recombinant virus. Properties that such a recombinant would have are unknown, but there is a potential homology between retroviral vectors and human T-cell leukaemia viruses so that the formation of a recombinant that could produce a malignancy is a possibility. There is, however, a built-in safety feature with the mouse retroviral vector now in use. These mouse structures have a very different sequence from known primate retroviruses, and there appears to be little or no homology between the two. Therefore, it should be possible, with continuing research, to build a safe retroviral vector.

With the present constructs, three types of experiments ought to be carried out before any retrovirus-treated bone marrow is injected into a patient. These protocols, designed to test the safety of the delivery-expression system, are necessary since once treated bone marrow is reinserted into a patient, it and all retroviruses that it contains are irretrievable.

First, studies with human bone marrow in tissue culture are needed. Marrow cultures infected with the therapeutic vector should be tested for a period of time for the production of recombinant viruses. Any infectious virus isolated should be studied for possible pathogenicity.

Second, studies *in vivo* with mice are needed. Treated animals should be followed to determine if genomic rearrangement or the site of chromosomal integration of the retroviral vector has resulted in any pathologic manifestations or the production of any infectious viruses.

Third, studies *in vivo* with primates are needed. A protocol similar to the one planned for human application should be carried out in primates, not just mice, because the endogenous viral sequences in primate, including human, DNA are different from those in mouse DNA. Therefore, the nature of any viral recombinants would be different. Treated bone marrow should be reimplanted into primates, the successful transfer of intact vector DNA into blood-forming cells demonstrated, the expression of at least small amounts of gene product verified, and the existence of infectious recombinant viruses sought.

Conclusion

It now appears that effective delivery expression systems are becoming available that will allow reasonable attempts at somatic cell gene therapy. The first clinical trials will probably be carried out within the next year. The initial protocols will be based on treatment of bone marrow cells with retroviral vectors carrying a normal gene. The safety of the procedures is the remaining major issue. Patients severely debilitated by having no normal copies of the gene that produces the enzyme HPRT, ADA or PNP are the most likely first candidates for gene therapy.

It is unrealistic to expect a complete cure from the initial attempts at gene therapy. Many patients who suffer from severe genetic diseases, as well as their families, are eager to participate in early clinical trials even if the likelihood is low that the original experiments will alleviate symptoms. However, for the protection of the patients (particularly since those with the most severe diseases and, therefore, the most ethically justifiable first candidates, are children) gene therapy trials should not be attempted until there are good animal data to suggest that some amelioration of the biochemical defect is likely. Then it would be necessary to weigh the potential risks to the patient, including the possibility of producing a pathological virus or a malignancy, against the anticipated benefits to be gained from the functional gene. This risk to benefit determination, a standard procedure for all clinical research protocols, would need to be carried out for each patient.

In summary, Institutional Review Boards and the NIH should carefully evaluate therapeutic protocols to ensure that the delivery system is effective, that sufficient expression can be obtained in bone marrow cultures and in laboratory animals to predict probable benefit, even if small, for the patient, and that safety protocols have demonstrated that the probability is low for the production of either a malignant cell or a harmful infectious retrovirus. Once these criteria are met, I maintain that it would be unethical to delay human trials. Patients with serious genetic diseases have little or no other hope at present for alleviation of their medical problems. Arguments that genetic engineering might someday be misused do not justify the needless perpetuation of human suffering that would result from an unnecessary delay in the clinical application of this potentially powerful therapeutic procedure.

154

Germ line gene therapy

The second level of genetic engineering, gene therapy of germ line cells, would require a major advance in our present state of knowledge. It would require that we learn how to insert a gene not only into the appropriate cells of the patient's body, but also how to introduce it into the germ line of the patient in such a way that it would be transmitted to offspring and would be functional in the correct way in the correct cells of the offspring. Based on the small amount of information available from animal studies, the step from correction of a disorder in somatic cells to correction of the germ line would be difficult.

Germ line therapy in animals

Germ line transmission and expression of inserted genes in mice has been obtained by several laboratories but with a technique that is not acceptable for use in human patients, namely the physical microinjection of fertilised eggs. Microinjection into tissue culture cells has been used for a number of years and has the advantage of high efficiency (up to one cell in five injected can be permanently transfected). However, the distinct disadvantage is that only one cell at a time can be injected. Transfection of a large number (like 10^6) of blood-forming stem cells is not feasible.

Microinjection has been used with considerable success in transferring genes into mouse zygotes. DNA can be microinjected into one of the two pronuclei of a recently fertilised mouse egg. This egg can then be placed into the oviduct of a pseudopregnant female where it can develop into a normal mouse carrying the exogenous DNA in every cell of its body including its germ cells. Consequently, the injected DNA can be transmitted to offspring in a normal Mendelian manner. Mice carrying an exogenous gene in their genome are called 'transgenic'.

It is this technique that was used to partially correct a mouse with a defect in its growth hormone production (Hammer *et al.*, 1984).[7] By attaching a rat growth hormone gene to an active regulatory sequence (specifically, the promoter that normally directs the synthesis of metallothionein messenger RNA in mice), researchers obtained a recombinant DNA construct that actively produced growth hormone in the genetically defective mouse and in a number of its offspring. Although the level of growth hormone

155

production was inappropriately controlled — that is, influenced by signals that normally regulate metallothionein synthesis — these experiments did show that microinjection can be used as a delivery system that can put a gene into every cell of an animal's body, that a genetic disorder can, as a result, be corrected, and that the correction can be passed on to the next generation of animals.

Why is the technique of microinjecting a fertilised egg not acceptable for use for human gene therapy at the present time? First, the procedure has a high failure rate; second, it can produce a deleterious result; and third, it would have limited usefulness. Microinjection has a high failure rate because the majority of eggs are so damaged by the microinjection and transfer procedures that they do not develop into live offspring. In one recent experiment (Brinster *et al.*, 1983)[8] involving microinjection of an immunoglobulin gene, 300 eggs were injected, 192 (64 per cent) were judged sufficiently healthy to be transferred to surrogate mothers, only 11 (3.7 per cent) proceeded to live birth and just 6 (2 per cent) carried the gene. These results are from a highly experienced laboratory in which thousands of identical eggs from the same hybrid cross of inbred mice have been injected over several years. The mice were chosen precisely because they gave the best results for gene transfer by microinjection. Attempts to microinject functional growth hormone genes into livestock eggs met with major biological and technical problems before being accomplished. Successful gene transfer by microinjection of human eggs, without a long period of trial and error experimentation, is extremely unlikely.

Second, microinjection of eggs can produce deleterious results because there is no control over where the injected DNA will integrate in the genome. For example, the integration of an exogenous rabbit beta-globin gene in transgenic mice can sometimes occur at a chromosomal location that results in expression of the beta-globin gene in an inappropriate tissue, viz., muscle or testis (Lacy *et al.*, 1983).[9] There have also been several cases reported where integration of microinjected DNA has resulted in a pathological condition. Although there is no control over where exogenous DNA will integrate in any gene transfer procedure, the damaging effect caused by a harmful insertion site could be great when it occurs in the egg but may be negligible when it occurs in one or a few of a large number of bone marrow cells.

The third objection to microinjection of eggs is limited usefulness. Not only is it ethically questionable to experiment on human eggs because of the expected losses, but even if 'success' were obtained, it would be applicable primarily when both patients are homozygous for the defect. When the parents are both carriers of a recessive trait, only one fertilised egg out of four would result in an affected child. Since a homozygous defect cannot yet be recognised in early embryos, and since the procedure itself carries such a high risk, it would be improper to attempt any manipulation in this situation. Furthermore, most of the very serious genetic disorders result in infertility (or death before reproductive age) in homozygous patients. Consequently, there would be little use for the procedure even if it were feasible.

Ethics

Even when the technical capability becomes available to attempt germ line gene therapy in humans, there are major medical and ethical concerns to consider. The medical issues centre primarily around the question: will the transmitted gene itself, or any side effects caused by its presence, adversely affect the immediate off-spring or their descendants? Since in this case one must study several generations of progeny to obtain answers, it will clearly take longer to gain knowledge from animal studies on the long-term safety of germ line therapy than on somatic cell gene therapy.

Germ line therapy deserves careful ethical consideration well in advance of the time when the technical capability for carrying it out arrives. The critical ethical question is: should a treatment which produces an inherited change, and could therefore per-petuate in future generations any mistake or unanticipated problems resulting from the therapy, ever be undertaken?

What criteria would be needed to justify the use of this unique type of therapy? At least three conditions should be met prior to the time that germ line gene therapy is attempted in human beings.

First, there should be considerable previous experience with somatic cell gene therapy that clearly establishes the effectiveness and safety of treatment of somatic cells. There is a wide range of biological variability among humans. Even if the first few patients treated by somatic cell therapy are helped, the next ones may not be, or may even be harmed. Therefore, extensive experience with many patients over a number of years will be necessary before somatic cell therapy can be properly judged to be safe and effective. If somatic cell therapy has not become highly efficient with minimal risks, then germ line therapy should not be considered.

Second, there should be adequate animal studies that establish the reproducibility, reliability, and safety of germ line therapy, using the same vectors and procedures that would be used in humans. Of greatest importance would be the demonstration that the new DNA could be inserted exactly as predicted and that it would be expressed in the appropriate tissues and at the appropriate times. It should be remembered that gene therapy does not remove or correct the defective genes in the recipient; it only adds a normal gene into the genome. It is not now known what the influence of this combination of defective and normal genes may be on the developing embryo. Might the regulatory signals still associated with the non-functional genes adversely affect the regulation of the exogenous gene during development?

Third, there should be public awareness and approval of the procedure. New drugs, medical regimens, and surgical techniques certainly do not require individual public approval prior to their initiation. There are already regulatory processes in place that ensure the protection of human subjects (this issue has been addressed in a previous publication (Anderson and Fletcher, 1980)).[10] Somatic cell gene therapy is receiving widespread public attention, but prior public approval is not being specifically sought. Germ line gene therapy, however, is a different and unique form of treatment. It will affect unborn generations and has, therefore, a greater impact on society as a whole than treatment confined to a single individual. The gene pool is a joint possession of all society. Since germ line gene therapy will affect the gene pool, the public should have a thorough understanding of the implications of this form of treatment. Only when an informed public has indicated its support, by the various avenues open for society to express its views, should clinical trials begin. *In vitro* fertilisation, surrogate motherhood, animal organ transplants into humans, holistic treatment of cancer, and other controversial medical procedures can take place based on the decision of the patient (with his/her doctor and/or family), whether society as a whole approves or not. But the decision to initiate germ line gene therapy demands assent from more than the individual involved, since the effects go beyond that individual. If and when germ line therapy is approved by society for clinical trials, then the decision to apply it in any individual case again should be made privately by the patient with his/her doctor.

In conclusion, my position is that germ line therapy, since it is the correction of a genetic defect (albeit in the future), would be ethical and appropriate if the three conditions discussed above were met.

Enhancement genetic engineering

The third level of genetic engineering, enhancement genetic engineering, is considerably different in principle from the first two. This is no longer therapy of a genetic disorder, it is the insertion of an additional normal gene (or a gene modified in a specified way) to produce a change in some characteristic that the individual wants. Enhancement would involve the insertion of a single gene, or a small number of genes, that code for a product (or products) that would produce the desired effect; for example, greater size through the insertion of an additional growth hormone gene into the cells of an infant. Enhancement genetic engineering presents a major scientific hurdle as well as serious new ethical issues. Except under very specific circumstances as detailed below, genetic engineering should not be used for enhancement purposes.

Scientific and ethical concerns

The scientific hurdle to be overcome is a formidable one. Until now, we have considered the correction of a defect, or a 'broken part', if you will. Fix the broken part and the human machine should operate correctly again. Replacing a faulty part is different from trying to add something new to a normally functioning system. To insert a gene in the hope of improving or selectively altering a characteristic might endanger the overall metabolic balance of the individual cells as well as of the entire body. Medicine is a very inexact science. Every year new hormones, new regulators, and new pathways are discovered. There are clearly many more to be discovered. Most impressive is the enormously intricate way that each cell co-ordinates within itself all of its thousands of pathways. Likewise, the body as a whole carefully monitors and balances a multitude of physiological systems. Much additional research will be required to elucidate the effects of altering one or more major pathways in a cell. To correct a faulty gene is probably not going to be dangerous, but intentionally to insert a gene to make more of one product might adversely affect numerous other biochemical pathways.

We possess insufficient information at present to understand the effects of attempts to alter the genetic machinery of a human. Is it wise, safe, or ethical for parents to give, for example, growth hormone (now that it is available in large amounts) to their normal sons in order to produce very large football or basketball players? Unfortunately, this practice now takes place in the USA. But even worse, why would anyone want to insert a growth hormone gene into a normal child? Once it is in, there is no way to get it back out. The child's reflexes, co-ordination, and balance might all be grossly affected. In addition, even more serious questions can be asked: might one alter the regulatory pathways of cells, inadvertently affecting cell division or other properties? In short, we know too little about the human body to chance inserting a gene designed for 'improvement' into a normal healthy person.

An acceptable use

There is, however, a set of circumstances under which enhancement genetic engineering may be ethical, in my opinion. This is when it could be justified on grounds of preventive medicine. For example, it is well established that heart attacks and strokes are a direct result of atherosclerosis (i.e. hardening of the arteries). The rate of development of atherosclerosis appears to correlate directly with elevated levels of cholesterol in the blood. The level of blood cholesterol is regulated, at least in part, by its rate of clearance from the blood by the low density lipoprotein (LDL) receptors on body cells (Goldstein and Brown, 1983).[11] LDL is the major cholesterol-transport protein in human plasma. If further research could verify that an increased number of LDL receptors on cells would result in lower blood cholesterol levels and, consequently, in a decreased incidence of heart attacks and strokes, then the insertion of an additional LDL receptor gene in 'normal' individuals could significantly decrease the morbidity and mortality caused by atherosclerosis. In this type of situation, the purpose of the intervention would be the prevention of disease, not simply the personal desire of the patient for an altered characteristic. The concerns expressed above about disrupting the regulatory pathway in the body still should be considered, of course. However, since there is a range for the number of receptors on a cell's surface, shifting a person with a 'low normal' number of receptors to a 'high normal' number may not be disruptive to other physiological or biochemical pathways.

Eugenic genetic engineering

The fourth level is 'eugenic' genetic engineering. This area has received considerable attention in the popular press, with the result that at times unjustified fears have been produced because of claims that scientists might soon be able to re-make human beings. In fact, however, such traits as personality, character, formation of body organs, fertility, intelligence, physical, mental, and emotional characteristics, etc., are enormously complex. Dozens, perhaps hundreds, of unknown genes that interact in totally unknown ways probably contribute to each such trait. Environmental influences also interact with these genetic backgrounds in poorly understood ways. With time, as more is learned about each of these complex traits, individual genes will be discovered that play specific roles. Undoubtedly, disorders will be recognised that are caused by defects in these genes. Then, somatic cell gene therapy could be employed to correct the defect. But the concept of 're-making a human' (i.e. eugenic genetic engineering) is not realistic at present.

Complex polygenic traits may never be influenced in a predictable manner by genetic engineering but, at a minimum, developing the techniques for producing such changes will take many years. Therefore, there is no point to a scientific discussion of eugenic genetic engineering at present — there is simply no science to discuss. But from a philosophical standpoint, a discussion of the ethics of eugenic genetic engineering is very important. After all, what is it that makes us human? Why are we what we are? Are there genes which are indeed 'human' genes? If we were to alter one of these genes, would we be other than human? These are important questions for us to think about and discuss.

If eugenic genetic engineering were possible today, I would be strongly opposed to its use on philosophical and ethical grounds. Our knowledge of how the human body works is still elementary. Our understanding of how the mind, both conscious and subconscious, functions is even more rudimentary. The genetic basis for instinctual behaviour is largely unknown. Our disagreements about what constitutes 'humanhood' are notorious. And our insight into what, and to what extent, genetic components might play a role in what we comprehend as our 'spiritual' side is almost non-existent. We simply should not meddle in areas where we are so ignorant. Regardless how fast our technological abilities increase, there should be no attempt to manipulate, for other than therapeutic reasons, the genetic framework (i.e. the genome) of human beings.

Conclusion

In summary, somatic cell gene therapy for human genetic disease should be possible in the very near future. The scientific basis on which this new therapeutic approach is founded has been thoroughly documented in a number of publications, as has the ethical justification for its use. Germ line gene therapy is still in the future, but the technical ability to carry it out will almost certainly be developed. Society must determine if this therapeutic option should be used. Enhancement genetic engineering should also be possible and its medical and disturbing ethical implications need continuing discussion. Eugenic genetic engineering, on the other hand, is purely theoretical and will be, from a practical standpoint, impossible for the foreseeable future. The topic is valuable for reflective thinking but not for scientific discussion.

Many of the fears generated by some articles in the popular press that discuss 'gene therapy' or 'genetic engineering' are simply unfounded. Insertion of single functional genes should soon become possible, but claims that new organs, designed personalities, master races, or Frankenstein monsters will be created can be given no credence in the light of what is presently known. Even so, we should be concerned about the possibility that genetic engineering might be misused in the future. The best insurance against possible abuse is a well-informed public. Gene therapy has the potential for producing tremendous good by reducing the suffering and death caused by genetic diseases. We can look forward to the day when, with proper safeguards imposed by society, this powerful new therapeutic procedure is available.

Notes and references

1. W. F. Anderson, 'Prospects for human gene therapy', *Science*, vol. 226 (1984), pp. 401–9.

2. J. C. Fletcher, 'Ethical issues in and beyond prospective clinical trials of human gene therapy', *Journal of Medicine and Philosophy* (1985), p. xxx.

3. W. F. Anderson and J. C. Fletcher, 'Gene therapy in human beings: when is it ethical to begin?' *New England Journal of Medicine*, vol. 303 (1980), pp. 1293–7.

4. Anderson, 'Prospects for human gene therapy'.

5. A. Joyner, G. Keller, R. A. Phillips and A. Bernstein, 'Retrovirus transfer of a bacterial gene into mouse haematopoietic progenitor cells', *Nature*, vol. 305 (1983), pp. 556–8; D. A. Williams, I. R. Lemischka, D. G. Nathan and R. C. Mulligan, 'Introduction of new genetic material into pluripotent haematopoietic stem cells of the mouse', *Nature*, vol. 310 (1984), pp. 476–80.

6. A. D. Miller, R. J. Eckner, D. J. Jolly, T. Friedmann and I. M. Verma, 'Expression of a retrovirus encoding human HPRT in mice', *Science*, vol. 225 (1984), pp. 630–2; R. C. Willis, D. J. Jolly, A. D. Miller, M. M. Plent, A. C. Esty, P. J. Anderson, H-C. Chang, O. W. Jones, J. E. Seegmiller and T. Friedmann, 'Partial phenotypic correction of human Lesch-Nyhan (hypoxanthine-guanine phosphoribosyltransferase-deficient) Lymphoblasts with a transmissible retroviral vector', *Journal of Biological Chemistry*, vol. 259 (1984), pp. 7842–9.

7. R. E. Hammer, R. D. Palmiter and R. L. Brinster, 'Partial correction of murine hereditary growth disorder by germ-line incorporation of a new gene', *Nature*, vol. 311 (1984), pp. 65–7.

8. R. L. Brinster, K. A. Ritchie, R. E. Hammer, R. L. O'Brien, B. Arp and U. Storb, 'Expression of a microinjected immunoglobulin gene in the spleen of transgenic mice', *Nature*, vol. 306 (1983), pp. 332–6.

9. E. Lacy, S. Roberts, E. P. Evans, M. D. Burtenshaw and F. D. Costantini, 'A foreign B-globin gene in transgenic mice: Integration at abnormal chromosomal positions and expression in inappropriate tissues', *Cell*, vol. 34 (1983), pp. 343–58.

10. Anderson and Fletcher, 'Gene therapy in human beings'.

11. J. L. Goldstein and M. S. Brown, 'Familial hypercholesterolemia' in J. B. Stanbury, J. B. Wyngaarden, D. S. Fredrickson, J. L. Goldstein and M. S. Brown (eds), *The metabolic basis of inherited disease*, (5th edn, McGraw-Hill, New York, 1983), pp. 672–712.

9

Eugenics on the Rise:
A Report from Singapore

C. K. Chan

Singapore's National Day in August 1983 was the occasion for a remarkably explicit statement by the Prime Minister, Lee Kuan Yew, of his firm convictions on intelligence, heredity, and their implications for social policy in Singapore.[1] For those of us who might have thought that vulgar eugenics had fallen into irredeemable disrepute in the hands of Nazi Germany's race hygienists, the recent developments in Singapore are a startling reminder that eugenic doctrines, in their more benign forms, retain much of their plausibility for large sectors of public opinion. Indeed, the appeal seems not to be confined to politically conservative societies, but is apparent even in the People's Republic of China, where there have been recent calls for 'positive eugenics . . . to accelerate the birth of individuals possessing outstanding physical and intellectual qualities'.[2]

In his National Day speech that year, Premier Lee returned to a favourite theme, i.e. the importance of the quality of human material that has sustained Singapore's impressive economic growth.[3] Coupled with his firm belief in the innate (indeed, hereditary) character of those qualities making for this success, eugenic considerations had been for quite a while an important factor in major areas of social policy in Singapore. In previous statements, public as well as private, Lee had been utterly explicit about his uncompromising elitism and his conviction that the fate of Singapore rested with an elite group of 'no more than 5 per cent' who were 'more than ordinarily endowed physically and mentally'.[4] This was not merely ideological cant, but was consistently carried over into social policy and most vividly typified by an elaborate tracking (streaming) system in the republic's educational set-up.

164

For Lee, the social engineering objectives were clear:

> We must expend our limited and slender resources [on these
> naturally superior individuals] in order that they will provide
> that yeast, that ferment, that catalyst in our society which
> alone will ensure that Singapore shall maintain its pre-
> eminent place in the societies that exist in South East Asia.[5]

In 1969, in a parliamentary debate on the Abortion Bill, Lee
had first expressed his concern for the undiminished propagation
of these superior individuals in the following terms:

> By introducing the new abortion law together with the com-
> panion voluntary sterilisation law, we are making possible the
> exercise of voluntary choice. But we must keep a close watch
> on the result of the new laws and the patterns of use which will
> emerge . . . One of the crucial yardsticks by which we shall
> have to judge the results . . . will be whether it tends to raise
> or lower the total quality of our population.[6]

By 1982, Lee's fears had begun to materialise in the form of a
widening differential between the fertility of graduate women
versus that of nongraduate women — a higher proportion of
graduate women were remaining single, and of those who
married, were producing fewer offspring on average than their less
educated counterparts. In Lee's view, these trends, if left
unchecked, could only lead to a dilution of human talent in
Singapore, a prelude to the certain demise of the island's hitherto
vibrant economy.

'Positive' eugenics: of computer dates, love-boat cruises and courtship classes

To underline the government's seriousness in this matter, Dr Goh
Keng Swee (Deputy Prime Minister at the time) unveiled to the
Singapore public a package of countermeasures to reverse these
trends. These included: a computer dating service; fiscal and other
incentives for graduate women to bear more children; love-boat
cruises (all expenses paid) for eligible graduate singles in the civil
service; special admissions criteria to the National University of
Singapore (NUS) to even out the male-female student ratio; calls

to NUS academicians to investigate the single graduate problem, and also the introduction of courtship classes in the undergraduate curriculum to hone the would-be suitor's skills, etc.

Predictably enough, such Orwellian schemes provoked quite an outcry, as it became clear that this was no laughing matter, and the government was entirely serious about launching a sustained campaign to achieve its demographic objectives. None the less, a quick reading of the popular reaction indicated that much of it was a response to the government's rather vulgar, technocratic approach to a matter of some human sensitivity. In other instances, graduate professional women wrote in to complain of problems in coping with career versus family responsibilities, of uncooperative attitudes among Singaporean males towards domestic duties, etc., quite legitimate concerns in and of themselves. But the basic premises, that intelligence (and by extension, social and professional success) was largely determined by genes, and the eugenic undertones of the measures being proposed, went largely unchallenged.[7] To be sure, there were scattered protests here and there, that bright parents did not necessarily beget bright children, but these seemed like layman sentiments against the weight of expert opinion cited by the Prime Minister:

> There is increasing evidence that nature, or what is inherited, is the greater determinant of a person's performance than nurture . . . Researches on identical twins who were given away at birth to different families of different social, economic classes show that their performance is very close although their environments are different. One such research, carried out for over a decade, is by Professor Thomas Bouchard of the University of Minnesota . . . The conclusions the researchers draw is that 80 per cent is nature, or inherited, and 20 per cent reflects the differences from different environment and upbringing.[8]

In the weeks that followed, Hans Eysenck, a prominent hereditarian in the IQ-heredity debate, was interviewed by a Singapore Broadcasting Corporation television team specially flown over to London for the purpose; Wong Hock Boon, professor of paediatrics of NUS and director of the WHO Regional Centre for Human Genetics, collaborated on a four-part series for the Singapore *Straits Times*, providing a popular account of genes and intelligence largely culled from the writings of Eysenck.[9] The regional press also chipped in, with a cover issue from *ASIA WEEK* featuring lengthy interviews with A. R. Jensen, William Shockley and Thomas Bouchard.[10] But the most crudely apologetic statement must have been that due to Dr Chou Kuan Hon, from the NUS Botany Department, who made the widely publicised assertion that ancient Greek civilisation declined because their upper-class women would not contribute their share of offspring. He further urged that Singapore should adopt a 'Selective Population Control' policy which would 'discourage birth control among highly intelligent women, but encourage it among the less intelligent ones'.[11] Dr Chou must have been gratified when the government announced

> a new family planning message [which] will tell different things to different people . . . [The new policy] encourages graduate women to have more children [for example, by guaranteeing their children placement in the republic's elite schools], but rewards less well-educated women who keep their families small.[12]

Anyone subjected to such a media barrage could hardly avoid the impression that a high degree of consensus existed, within scientific circles, and that the predominant role of genes in IQ determination was a demonstrated truth. This was an impression so much at variance with the current state of the debate that the conspicuous silence of local academics and professionals (excepting Professor Wong and Dr Chou) was most unfortunate in reinforcing this erroneous picture. This was all the more so in light of the devastating critiques, in the 1970s, of practically the entire body of empirical data as well as the theoretical edifice which had been offered in support of the hereditarian position.

The 'scientific' status of IQ research

The scientific status of IQ research became an issue of vigorous debate in the 1970s, following A. R. Jensen's famous article in the *Harvard Educational Review* ('How much can we boost IQ and scholastic achievement?').[13] In the early exchanges, much was made of the observed high heritabilities of IQ (70–80 per cent), the very same figure that Lee constantly referred to. In particular, these figures relied heavily upon four studies of separated identical twins, most importantly those by the late Sir Cyril Burt, who reported the largest sample size, and whose experimental design up until 1974 was regarded as the most rigorous. In that year, Leon Kamin published the results of an extensive re-examination of the empirical data on IQ and genetics (under the title *The science and politics of IQ*),[14] in which he showed strong circumstantial evidence of data fabrication by Burt. Over the next five years, additional information came to light, further substantiating Kamin's suspicions. Finally, in 1979, Burt's scientific fraud was definitively confirmed by his biographer, Leslie Hearnshaw, who had been commissioned by Burt's sister and had been given access to his confidential papers, diaries and correspondence.[15]

Kamin's review, however, was not confined to Burt's studies. For the remaining twin studies, and also the kinship and adoption studies, he showed major flaws in experimentation, analysis and interpretation, such that by the end of his review, he comes to the conclusion that 'there exists no data which should lead a prudent man to accept the hypothesis that IQ test scores are in any degree hereditable'.[16]

Quite apart from the discredited empirical data, the entire theoretical framework of heritability analysis has been subjected to severe criticism, in particular by R. C. Lewontin;[17] for instance, the heritability measure of a human trait (e.g. the alleged 80 per cent for IQ) is a value that can change depending on the range of environments in which the trait was measured. Furthermore, a high heritability for a given trait does not imply the immutability of that trait.

To be sure, critics like Kamin and Lewontin do not claim to have proved that genes have nothing to do with human intelligence. What they do say is that the evidence, as it exists, in no way justifies the claims of Eysenck or Jensen, and most certainly provides no basis whatsoever for the kinds of policies being implemented in Singapore.

'Negative' eugenics: the Shockley plan for Singapore

Notwithstanding the intense publicity and efforts at enlisting scientific endorsement, Lee's technocrats soon discovered that it would take more than tax rebates and preferential school placement to induce graduate women to produce more offspring. The complementary measure, to encourage lesser-educated women to have smaller families, was duly announced on 2 June 1984. Under this scheme, women who met certain conditions would be rewarded with a lump sum of US$ 4,000 as down payment for a government low-cost apartment. These conditions were:

below thirty years of age
have two or less children
educational level not beyond junior high school
neither she nor her spouse earning more than US$ 300 a month,
and were willing to be sterilised.

This proposal, sensitive now to the sentiments which had been aroused, was phrased in the language of an anti-poverty measure, in effect playing upon the needy circumstances of the low-income people.[18]

The initial response from the target group took the government rather by surprise, as calls started flooding in from eligible couples. None the less, this pragmatism belied a simmering resentment, which was to become fully evident six months later during the republic's general elections. Opposition parties picked on this issue, among other issues, and apparently the disaffection ran deep enough to result in a substantial swing of the popular vote away from Lee's People's Action Party (PAP), a party that had been accustomed until three years ago to a total control of the Singapore Parliament. In the wake of this electoral setback, the PAP has back-pedalled on some of its more controversial schemes: the incentives for graduate mothers were dropped and the tracking system in the schools was to be thoroughly reviewed. For the moment however the incentives for sterilisation, probably the most 'successful' among the eugenic measures, remained on the books.

Observers of the Singaporean scene have often marvelled at the degree of autonomy from sectarian interests, with which the technocracy seems to function. This is in contrast to many other capitalist countries in which short-term, factional interests manage to exert much more influence upon state policies, in the extreme even threatening to drive the whole system to collapse. The Singapore technocracy, on the contrary, has repeatedly shown itself quite willing and capable of undertaking those measures which it believes to be in the long-term interest and viability of the system as a whole. So for instance, in line with its policy of technological upgrading (to maintain its competitive edge over its neighbours in the region), the government mandated hefty wage increases, ranging from 14 to 20 per cent annually, for three consecutive years from 1979 to 1981. The intention was to encourage the phasing out of labour-intensive industries (such as textiles), but it none the less raised a veritable storm in many a corporate board-room. Not the kind of policy that one would normally expect of a capitalist state, but neither is it suggestive of a pro-labour stance. Similarly, the introduction of the eugenic measure, for the social planners, merely expresses a concern that the 'quality' of the Singaporean population measures up to the demands of an advanced, highly technological society.

There is, of course, another level of interpretation, which speaks to the ideological role of biological determinist perspectives in modern class society.[19] In the United States, for instance, one can readily appreciate that the views of Jensen, Herrnstein,[20] or the sociobiology school would be tremendously appealing to the privi-leged strata of society, given the ideological ambience brought on by the demands of the black liberation and feminist movements in particular. Undoubtedly, there are also many in Singapore who similarly hold that disparities between class, gender and ethnic groups are unavoidable manifestations of underlying biological differences. In the particular situation of Singapore, however, this ideological expression of privileged class interest emanates not so much from an overt challenge to a racist and patriarchal bourgeois society, but has been mediated and much amplified by the vigorous initiatives of a remarkably technocratic capitalist state.

Notes and references

1. K. Y. Lee, National Day Speech (14 August 1983). Reprinted as 'The education of women and patterns of procreation', in *RIHED (Regional Institute of Higher Education and Development) Bulletin*, vol. 10, no. 3 (1983).

2. R. Yuan, 'Paying great attention to eugenics and improving the quality of population', *Renkou Yanjiu [Population Research]*, vol. 3 (1983), translated in *Population and Development Review*, vol. 9, no. 4 (1983), pp. 756-61.

3. Singapore, together with South Korea, Taiwan and Hong Kong have been acclaimed as the most successful among the newly industrialis-ing countries, the four 'little dragons' whose rapid and sustained growth over the last decade and a half has propelled them almost into the ranks of the industrialised countries. For a discussion of the Singapore case see L. Lim, 'Singapore as No. 1', in Jomo (ed.), *The sun also sets* (INSAN, Kuala Lumpur, 1985).

4. M. Caldwell, *Lee Kuan Yew: the man, his mayoralty, and his mafia*

(Federation of UK and Eire Malaysian and Singaporean Student Organizations, London, 1979), p. 13.

5. Quoted in Caldwell, *Lee Kuan Yew*.

6. K. Y. Lee, *Abortion Bill, Third Reading* (as reported by a Parliamentary Select Committee), 29 Dec. 1969, p. 323.

7. Indeed, there was a substantial bedrock of public opinion that agreed with Lee, and even encouraged such measures as a tax on the 'well-educated' men in their thirties who wished to remain single, quite apart from all those other measures to induce graduate women to produce more offspring.

8. K. Y. Lee, National Day Speech, 1983, p. 4.

9. *Straits Times*, 7 Sept. 1983; 9 Sept. 1983; 11 Sept. 1983; 14 Sept. 1983.

10. *ASIAWEEK*, 2 March 1984.

11. *Straits Times*, 13 Sept. 1983.

12. *Straits Times*, 31 Jan. 1984.

13. A. R. Jensen, 'How much can we boost IQ and scholastic achievement?', *Harvard Educational Review*, vol. 39 (1969), pp. 1–123.

14. L. J. Kamin, *The science and politics of IQ* (Penguin, Harmondsworth, 1978).

15. L. Hearnshaw, *Cyril Burt: psychologist* (Cornell University Press, Ithaca, 1979).

16. Kamin, *Science and politics of IQ*, p. 15.

17. R. C. Lewontin, 'The analysis of variance and the analysis of causes', *American Journal of Human Genetics*, vol. 26 (1974), pp. 400–11; M. W. Feldman and R. C. Lewontin, 'The heritability hang-up', *Science*, vol. 190 (1975), pp. 1163–8.

18. Statement from the Prime Minister's Office, 2 June 1984, Singapore.

19. S. Rose, L. J. Kamin and R. C. Lewontin, *Not in our genes: Ideology, biology, and human nature* (Penguin, Harmondsworth, 1984).

20. R. J. Herrnstein, *IQ in the meritocracy* (Little Brown, Boston, 1973).

10

Should One be Free to Choose the Sex of One's Child?

Dharma Kumar

It is now possible to discover the sex of a foetus.[1] Clinics special-ising in these tests have been set up in India, and on the basis of their predictions female foetuses in particular are aborted. This has led to widespread disquiet in India, and to demands by feminists and others that the clinics should be banned, although many hold at the same time that there should be no restrictions on abortion itself. (We may call this the 'feminist' view for convenience, though not all feminists hold it, and some non-feminists do.) The Indian Health Minister expressed government disapproval of the.clinics in Parliament (on 26 July 1982) but they are still legal. Whether or not such tests should be banned raises several ethical issues, some of which are taken up in this paper.

The tests raise moral issues because they may provide the basis for decisions on whether to abort the foetus or not.[2] For instance, people who want only boys will abort all girl foetuses. One's views on the tests will therefore be connected with one's views on abortion, but not in obvious ways.

We may distinguish between three classes of people.[3] First, there are strict anti-abortionists, i.e. those who believe that abortion is always wrong, perhaps because the foetus has an abso-lute right to life. The second group, 'modified anti-abortionists', believe that abortion is wrong in itself but may be justifiable under certain specified, limited circumstances, such as the health of the mother, or the likelihood of grave disease. Finally, 'neutralists' feel that there is nothing wrong in the act of abortion itself. Never-theless some neutralists may hold that unlimited free choice in the matter of abortion may have undesirable social consequences and that social policies restricting individual free choice will therefore be justified. Such people may even feel that under certain circumstances, such as the likelihood of disease or overpopulation, abortion should be encouraged, or even enforced; while in other circumstances it should be discouraged or prevented.

Strict anti-abortionists

Those who hold that abortion is always wrong will object to any prediction of future characteristics if it increases the likelihood of abortions. This will depend on how likely it is that the characteristic being tested for will occur and how strongly it is disliked. For instance, even if parents have strong objections to a mongol child, if the probability of having such a child is very low, and if no test for this characteristic is available, they will risk having the child. But when it is possible to test for this condition, foetuses suffering from it will be aborted. So when the test is made to discover the presence of some unwanted characteristic, such as congenital idiocy, on the basis of which the decision to abort or not will be made, the availability of knowledge of the characteristics of a foetus will generally increase the chances of its abortion. On the other hand, the availability of the test may not, in fact, increase abortion, if the characteristic being tested for is highly undesirable or if its likelihood is high. Where either parent is known to be a potential carrier of some dreaded disease, in the absence of such knowledge they may prefer to abort all foetuses rather than take the risk of producing a child suffering from the disease. Muscular dystrophy is known to be genetically transmitted but only male children are affected. Parents who formerly preferred to have no child to taking the risk of a diseased one can now test for sex and abort only male foetuses. Tests for sex will reduce the likelihood of abortion in such cases, and it follows that a strict anti-abortionist need not necessarily disapprove of them (note that the test may be for sex, not for the disease itself). To be more precise, if one cannot test directly for a disease, and if the susceptibility to the disease is sex-specific, then tests for sex may actually reduce the likelihood of abortion. This argument holds even when there is no question of disease. The likelihood of a foetus being female is high, in fact 50 per cent, so if the mother greatly dislikes the prospect of a daughter and cannot predict the sex of the child, she will abort rather than take the risk of having a daughter. If she can predict the sex of a child, the likelihood of abortions is halved.

However, one may hold that contraception is always better than abortion and that the availability of tests for sex will encourage such parents to conceive, and then abort the unwanted foetus. Moreover, a strict anti-abortionist may feel that calculating the chances of abortion and then choosing the policy that minimises the probability of abortion, is beside the point — any act undertaken with a view to possible abortion is wrong and hence tests should be allowed only when abortion is not in question.

Modified anti-abortionists

The modified anti-abortionist holds that the foetus has a right to life, but not an absolute right. Its rights may be overridden by the right of others, the least controversial example being the right of the mother to life. Or one might argue that the right to life belongs only to those foetuses which will develop into children with a minimum level of physical or mental capacity.

Another widely accepted ground for abortion is potential disease; modified anti-abortionists will therefore approve of tests for sex in the case of sex-related diseases, since they increase the likelihood of abortion of foetuses subject to disease. But sex in itself is hardly likely to be considered a sufficient ground for abortion so, apart from this minor exception, they, like the strict anti-abortionist, should consider whether the availability of the tests will increase the likelihood of abortions or not.

The neutralist position

Tests for sex pose a dilemma to those who do not believe that the foetus has an inherent right to life. If one were to consider the tests only as a question of private decisions, one would see them as enlarging freedom of choice, so for this reason libertarians should approve of them, or at least not oppose them, though some may feel that the ability to plan the sex composition of one's family is trivial or misleading — parents will love the unwanted girl as much as the hoped-for boy. However, one may hold that if people prefer sons they will treat them better than daughters (or vice versa), so allowing them to choose between sons and daughters will increase the number of happy children in the world.

But what about the possibility that private decisions will lead to an undesirable social result, in particular to an undesirable ratio of men to women (the sex ratio)? Apparently in many, perhaps most, societies people would rather have sons than daughters. Etzioni quotes an estimate that if this preference for boys could be fulfilled with mass availability of sex determination tests in the US, men would exceed women by 9.5 out of 100. Perhaps it is too soon to check the accuracy of the forecast, but noting that 'most forms of social behavior are sex-correlated', Etzioni argues that 'a significant and cumulative male surplus will thus produce a society with some of the rougher features of a frontier town'.[4]

All kinds of other considerations could be advanced against an unbalanced sex ratio, for instance that laws to enforce equality of pay and so on will be less likely to be enacted when women are in a minority in the population. (Analogously, in multi-religious democracies like India, minorities are urged to have large families to add to the communities' vote banks.)

The underlying premiss of these arguments is that tests for sex will bring about an overall fall in the general welfare over the short or long run, which is greater than the increase in the welfare of parents and children in families which choose between boys and girls. The little boy may be happier than the little girl would have been, but this difference in happiness is much less than his unhappiness in a world where he cannot find a wife, where other men are aggressive, and so on, not to mention the unhappiness of other people. Of two societies with equal numbers, the one where people can choose the sex of the child, and end up with far more boys than girls, is less happy than one where they cannot choose, and which therefore has a more balanced population.[5] Many people therefore draw the conclusion that individual free choice in the matter of the sex of one's children is not desirable, and hence that policies to ban tests or discourage them are desirable. Analogous policies spring to mind — one could argue that suppressing information about the sex of the foetus is like not telling a judge about the irrelevant past of a defendant, or like keeping information secret from a potential employer who is irrationally prejudiced about sex.

India and China

These problems are posed most sharply in some non-Western societies where, first, the parents' preference for boys over girls

175

is much stronger than in the West and there is hence a greater danger of a surplus of males resulting from the mass availability of tests. Secondly, this preference is so strong that girls are often neglected, ill-treated or killed. And finally, in traditional as well as communist societies, parents are not always free to take decisions about their own families.

To take the last point first, in many traditional societies the decision to abort is taken in consultation with, or even solely by, the elders of the joint family. Indian feminists have argued that these elders are likely to have a greater boy-preference than the couple, and that with the availability of tests they will insist on the abortion of the female foetus. This has two bad effects. The mother is forced to do something she does not want to do (and which may affect her health).[6] Secondly, the ratio of men to women in India, already too high, will increase even more.

A similar argument could be advanced with respect to Communist China. One effect of a shortage of women is that the rate of population growth is reduced, so that it is conceivable that if the government wants to cut population growth sharply, even mothers who want girls will be forced to abort female foetuses. In both these cases, availability of knowledge about the sex of the foetus may be used to restrict the mother's freedom. However 'feminists' urge that in order to prevent an unbalanced sex ratio the tests should be banned even if the mothers themselves prefer boys to girls as they often do. But one must also consider the fate of unwanted girls. There are two aspects to the preference for boys. First, while people want both sons and daughters, sons are considered essential and daughters are not, and, secondly, fewer daughters are desired than sons. Thus, a son is essential in the performance of certain rituals,[7] and in both India and China a son may be the only source of support in old age. On the other hand, daughters are expensive — they need dowries, and their labour is contributed to the husband's family.

If couples cannot predict the sex of the child, they will go on having children till they get at least one son; if this means that they have more daughters than they want, they may neglect and ill-treat them. There is extensive evidence of the neglect and ill-treatment of girls in India.[8] In fact, one can extend the argument to mothers — a woman who does not have sons may be ill-treated by her husband's family. 'Feminists' who are not opposed to foeticide as such, but only to sex-selective foeticide, may argue that the quality of individual female lives is less important than their relative numbers in the population, at least in the short run. It is conceivable that the higher the proportion of women in the population the more likely their condition is to improve over time, however ill-treated they are to start with, but this is not very plausible — ill-educated, weak girls and women are unlikely to exert much political power. And in fact one could as well argue the opposite case — that scarcity will improve the status of women (here the analogy with racial and religious minorities breaks down because women are essential to societies in ways in which ethnic minorities generally are not).

Ill-treatment of girls runs all the way from undernourishment to near-infanticide and outright infanticide. It is difficult to tell how widespread female infanticide is; the point is that it may well increase if amniocentesis and hence female foeticide are not possible, and the size of the family is restricted. It is reported that in China, the insistence on one-child families has resulted in one-fourth of girl babies being killed in some regions in the last few years. A recent report in *The Economist* states:

> Now the Anhui women's federation has surveyed the infant survival rates in two years. In 1980 and 1981 40 baby girls were drowned in one village. In another, in one three-month period of five new-born girls, three were drowned and two abandoned.
>
> In China, as elsewhere, slightly more boys are born than girls. In the same two areas, in 1979, the population that survived their first days reflects this; the survey found surviving babies split 51½ to 48½ per cent in favour of boys. But then the government's one-child family drive began. By 1981, the figures were 58 per cent boys, 42 per cent girls. The implication of these figures is hair-raising: whatever the 'natural' rate of infant mortality, roughly one female in four who might expect to survive fails to do so . . .
>
> If a peasant couple is expecting a child they want it to be a 'good' one — a boy. It is relatively easy to murder a new-born child in fact, as long as it is done within three days. Babies who die within that time are regarded as still-born and their births are not registered. Midwives in Guamgdon, according to local newspapers, plunge a girl baby straight into a bucket of water at the moment of birth and simply record a still birth. (*The Economist*, 16 April 1983)[9]

In such a situation, it seems difficult to make a case against tests for sex — one would have to argue that foeticide is morally worse than infanticide. Most people would hold the opposite view, that infanticide is worse, though this may reflect one's lack of imagination.

> If we could see the foetus from outside, would we still be so sure about the immense difference between the foetus-baby just before and just after birth? If we would not, this suggests that enthusiasm for this boundary is more the result of the foetus being out of sight than of any real 'onset of personality' at birth. This seems like the view that starvation and death matter far less in a faraway country, since we are less aware of it than if it were happening here.[10]

As Glover points out, this is to argue only that infanticide is no worse than foeticide, not that it is better. And even if it is granted that there is no difference between the two the choice should be left to the mother or parents.

The 'feminist' might still argue that the vast majority of people in fact feel that infanticide is morally worse than foeticide, and that making tests for sex available will therefore result in more female foetuses being destroyed than girl babies would have been without tests. The implicit argument is that what matters is keeping the total of female infanticide and female foeticide down because the proportion of women in the population must be maintained. But the proponents of this argument would have to admit that the ban may increase the probability of female infanticide — the Chinese example is proof of that. Infanticide is not much easier to detect and punish than foeticide, especially in a society where it is accepted. And on their own reasoning infanticide has worse side effects, since it is perceived as more wicked than foeticide.

One must therefore search for alternative policies to combat boy-preference.

Alternatives to banning tests

The choice of better policies will depend partly on the causes of the preference. If it is economic, one can try directly to increase the economic prospects of daughters or offset the losses from not having a son — the policies suggested in India include such measures as special employment measures for women and special pensions for couples without sons. Where the boy-preference is based on cultural or psychological factors, changing it may require different measures, such as education, whose results are partly unknowable. However, while it is well-known that causes will in fact be various and partly unknowable, it is well known also that purely economic measures can affect cultural preferences.

To solve a different problem, that of combining a social policy for birth control 'with a maximum of individual liberty and ethical choice', Boulding has suggested a system giving each girl approaching maturity a certificate entitling her to 2.2 children (or whatever number is needed for the desired rate of growth); certificates could be bought, sold or gifted, enabling individuals to indulge their preferences for large or small families within the overall total.[11] If this scheme is workable at all, it could easily be adapted to secure the desired sex ratio — each girl would be given a certificate for 1.1 boys and 1.1 girls; in societies with high boy-preference the price of the boy-certificates would be very high.

This scheme has ethical problems too (such as of distribution), some of which Boulding discusses, but its emphasis on the need to preserve individual liberty is very important.[12]

Conclusions

Tests for sex are now made after conception, and may lead to abortion — the question of justification arises in relation to the right to life of the foetus, and we have discussed the morality of the tests, given different views regarding this right: the foetus has an absolute right to life, a modified right, or no right. On the whole, those who believe that the foetus has some right to life are unlikely to favour tests for sex, except in the limited case of diseases which cannot be directly detected but are sex-specific. Even those who do not believe in the right to life of the foetus may hold that the availability of tests for sex will lead to undesirable social results.

But it may become possible to choose the sex of the child at the time of conception, so the question of the right to life of the foetus will then become irrelevant, leaving for consideration only the moral issues considered by those who hold that there is no such right. In particular, should parents be free to choose the sex of their children?

The case for banning this choice must proceed on the following lines. First, a society with a balanced sex ratio is happier than one with a large preponderance of one sex or another. However, since most people prefer boys to girls, if they are allowed to choose the sex of their child, an unbalanced sex ratio will result, so people must not be allowed to choose. The loss in individual freedom of choice and (short run) happiness of a few is outweighed by the unhappiness of an unbalanced society.

This argument might seem to be strongest where the preference for boys is strongest but there one must consider the fate of the girls who do survive. In countries like India and China, those making the case must argue that female foetuses are more likely to be destroyed than female babies (since babies are easier to love than foetuses), and that the surviving female children will not be so ill-treated as to make their lives not worth living. They must also argue that other methods of combating the preference for boys are not feasible, at least in the short run, and that banning tests is feasible. On balance, the case for banning tests — as proposed by 'feminists' in India — is weak.[13]

Notes and references

1. This is now done by amniocentesis, or extracting and analysing the fluid in the amniotic sac of a pregnant woman. This may involve some risk to the woman, but safer procedures are likely soon to be discovered, so the question of risk to the mother is ignored in this discussion. This is also one reason why we use the general term 'test for sex' rather than the unfamiliar 'amniocentesis'. The question of the feasibility of separating information on sex from information on other characteristics, such as liability to disease, is ignored.

2. Few people are likely to have moral objections to tests in themselves, though one conceivable religious objection to any test predicting the characteristics of a child is that one should not anticipate the will of God. Another possible objection has been pointed out by Etzioni — that the predictions may have harmful effects: for instance if parents are told that the child has an XYY chromosome structure which is associated with 'serious deviant behaviour', they may treat the child with suspicion (A. Etzioni, *Genetic fix: the next technological revolution* (Harper Colophon, New York, 1973), p. 195). These chromosomes are extremely rare, but if ways are discovered of predicting intelligence, however defined, the dangers of self-fulfilling prophecies may become serious — for instance, parents may neglect the education of children they have been told are not intelligent. The opposite is also possible — they may spend extra effort on such children, and in any case these considerations do not apply to tests for sex.

3. This is somewhat different from Tooley's threefold classification into liberals, who believe that abortion should always be permitted because

abortion is never wrong in itself and need not involve undesirable consequences; the anti-abortionists, who feel that it is almost never permissible, and moderates, who hold some position between (M. Tooley, *Abortion and infanticide* (Oxford University Press, Oxford, 1982), p. 285). Since we are interested primarily in attitudes to the test for sex, and in views on abortion only as determining them, the classification adopted seems preferable.

4. Etzioni, *Genetic fix*, p. 230. Apart from the early American settlements, on which there is an enormous literature, there have been several other societies with sex ratios unbalanced by migration, war, etc., and with different social problems and different ways of coping with them.

5. In fact there will be inequalities between the numbers of boys and girls born, but one can safely assume that these will not be very large. In any case, that raises a different moral problem — whether people should be forced to have children of a certain sex; here one is considering whether they should be prevented from choosing.

6. But the balance of power within the traditional family is changing, at least in some parts of the country. In a village study in south India it was found that over the last three generations the power to take decisions about the size and sex composition of the family had moved from the elders to the potential parents (J. Caldwell, P. T. Reddy and P. Caldwell, 'Demographic change in rural south India', *Population and Development Review*, vol. 8 (1982)).

7. It is true that adopted sons can perform the ritual, but traditional families were reluctant to adopt children from outside the family.

8. B. D. Miller, *The endangered sex: neglect of female children in rural north India* (Cornell University Press, Ithaca, 1981), summarises much of the literature on the ill-treatment of girls in India but her treatment is sometimes unbalanced; Dharma Kumar, 'Utopias and female nightmares', *Economic and Political Weekly*, vol. XVIII, no. 3 (15 Jan. 1983), pp. 61–4; R. Jeffrey and P. Jeffrey, 'Female infanticide and amniocentesis', *Economic and Political Weekly*, vol. XVIII, nos. 16–17 (16–23 April 1983), pp. 654–6. On the neglect of female children, see also J. Kynch and A. Sen, 'Indian women: wellbeing and survival', *Cambridge Journal of Economics*, vol. 7 (1983), pp. 363–80. The ill-treatment of unwanted children is not restricted to poor societies, as shown by the study of 120 Swedish children born after their mothers were refused abortion, quoted by Hardin in G. Hardin (ed.), *Population evaluation and birth control*, 2nd edn (W. H. Freeman, San Francisco, 1969), p. 298, but the ill-treatment is probably greater in poor societies.

9. The demographer, Ansley Coale, has calculated that around 250,000 baby girls have been killed in China since 1979. (*Economist*, 11–17 August 1984, p. 22). According to reports in Chinese newspapers, even girls as old as five years are killed (Minneapolis *Star and Tribune*, 17 April 1983).

10. J. Glover, *Causing death and saving lives* (Penguin, Harmondsworth, 1977), p. 126.

11. G. K. Boulding, 'Marketable licenses for babies' in Hardin, *Population evaluation*.

12. This need to preserve individual liberty is particularly strong in issues regarding total population; the increase in female infanticide in

China is a direct consequence of the limit of one child per family. The constant propaganda about overpopulation as the worst danger facing the world reaches madmen in authority in densely populated, poor and illiberal societies, with terrible consequences, as India under the Emergency and China show. And these usually meet with ready approval in the West, the Moral Majority being a bizarre exception. To those, like myself, who do regard overpopulation as a terrible danger, but who are very reluctant to accept the necessity of totalitarian measures to overcome it, allowing tests is better than coercive birth control. Reducing the number of girls slows down the birth-rate, and has the additional advantage of meeting the desires of the parents.

13. I am grateful for their comments to Sudhir Anand, John Broome, Radha Kumar, J. Krishnamurthy and Amartya Sen and for their hospitality to St Catherine's College, Oxford and the University of Warwick.

Glossary

AID: fertilisation of a woman's egg with semen from a donor, without sexual intercourse. Semen is placed in the woman's vagina.

Alleles: alternative forms of a gene which may occupy the same locus on each chromosome of a pair.

Chorion villus sampling: tissue is taken from the chorion, one of the membranes surrounding the foetus, with a view to checking its DNA. This can be done eight to ten weeks into pregnancy.

Chromosomes: the threads visible as pairs in body cells, carrying the genes in linear arrangement.

Clone: population of organisms derived from a single cell ancestor by repeated cell divisions.

Cytoplasm: the substance of the cell, as distinguished from the nucleus.

Ectogenesis: development of embryos outside the womb, e.g. in an artificial womb.

Egg donation: a process whereby a fertile woman donates an egg to be fertilised *in vitro* by the semen of the partner of an infertile woman.

Embryo donation: a process whereby a donated egg is fertilised by donated sperm *in vitro* and the resulting embryo transferred to a woman who is herself infertile and whose partner is infertile.

Endogenous genetic material: the genes belonging to an organism, as opposed to genes from an external source.

Episomes: genetic elements that can be added to the main chromosome of a cell.

Eugenics: suggestions for improvement of human genetic material.

Exogenous genetic material: genetic material from an external source.

Gametes: reproductive cells, sperm and egg. Each contains half the total number of chromosomes.

Gene: that element of DNA in which the amino-acid sequence of a protein is encoded.

Gene pool: the sum total of genes in the reproductive cells of a population.

Genome: the genetic complement of an organism taken as a whole.

183

Genotype: the genetic constitution of an individual.

Germ cell: reproductive cell.

GIFT: (gamete intra-fallopian tube transfer) a process which avoids laboratory fertilisation and implantation, by an operation which transfers the egg into the right position so that it can be fertilised *in vivo*.

Heterozygous: having different alleles at a gene locus.

Homozygous: having the same alleles at a gene locus.

In vitro **fertilisation (IVF):** fertilisation of a woman's egg by a man's sperm in the laboratory, by mixing them, not in a test tube but in a petri dish.

Karyotype: a description of the chromosome pairs possessed by a given cell type.

Messenger RNA: the single-stranded RNA copies of DNA which carry the information from the nucleus to the cytoplasm. The DNA is in the nucleus; the work of making protein goes on in the cytoplasm. Hence the need for a messenger.

Mutation: spontaneous or induced change in a gene.

Parthenogenesis: propagation without male, by prompting an egg to divide without the intervention of sperm.

Phage (bacteriophage): a virus which infects a bacterial cell.

Phenotype: overt character of an organism.

Polygenic trait: a trait controlled by a number of genes.

Recombinant DNA: i.e. 'genetic engineering': a process whereby DNA from one organism is introduced into the DNA of another organism e.g. the incorporation of a mammalian gene into the DNA of a bacterium.

Somatic cell: a body cell, not a reproductive one.

Surrogacy, full: the term used by Peter Singer and Deane Wells to describe a process in which the carrying mother does not contribute any genetic material. The egg of the female half of the commissioning couple is fertilised *in vitro* by the sperm of her male partner, and then transferred to the carrying mother.

Surrogacy, partial: where the carrying mother is also the genetic mother of the child. The carrying mother is either artificially inseminated or has intercourse with the would-be father.

Transcription: the process by which information carried in the DNA is converted into information encoded in messenger RNA.

Transduction: the carrying over of DNA from one organism to another by an intermediate agent, such as a virus.

Translation: the process by which the information carried by messenger RNA is decoded and expressed as a sequence of amino-acids.

Select Bibliography

Acton, H. B. and Watkins, J. W. N., 'Symposium: negative utilitarianism' in *Proceedings of the Aristotelian Society*, supplementary vol. 37 (1963).

Anderson, W. F., 'Prospects for human gene therapy', *Science*, vol. 226 (1984), pp. 401–9.

—— and Fletcher, J. C., 'Gene therapy in human beings: when is it ethical to begin?', *New England Journal of Medicine*, vol. 303 (1980), pp. 1293–7.

Annas, G. J., 'The case of Elizabeth Bouvia', *Hastings Center Report*, vol. 14, no. 2 (1984), pp. 20–3.

Baer, A. S. (ed.), *Heredity and society: readings in social genetics*, 2nd edn (Macmillan, New York, 1977).

Barnet, A., *The human species*, 3rd edn (Penguin, Harmondsworth, 1968).

Beckwith, J. and King, J., 'The XYY syndrome: a dangerous myth', *New Scientist*, 14 Nov. 1974, pp. 474–6.

Benjamin, M., Muyskens, J. and Saenger, P., 'Short children, anxious parents: is growth hormone the answer?', *Hastings Center Report*, vol. 14, no. 2 (1984), pp. 5–9.

Board for Social Responsibility, *Human fertilisation and embryology: the response of the Board for Social Responsibility of the General Synod of the Church of England to the DHSS Report of the Committee of Inquiry* (Ludo Press, London, 1984).

—— *Personal origins: the report of a working party on human fertilisation and embryology of the Board for Social Responsibility* (CIO Publishing, London, 1985).

Bromhall, D., 'The great cloning hoax', *New Statesman*, 2 June 1978.

Burnet, M., *Genes, dreams and realities* (Penguin, Harmondsworth, 1973).

Callahan, D. *et al.*, 'Ethical and scientific issues posed by human uses of molecular genetics' in Lappé, M. and Morrison, R. S. (eds), *Annals of the New York Academy of Sciences*, vol. 265 (1976).

Carter, C. O., *Human heredity* (Penguin, Harmondsworth, 1962).

Chadwick, R. F., 'The ethics of eugenics and genetic engineering' in *Ethics: foundations, problems and applications* (Proceedings of the 5th International Wittgenstein Symposium) (Holder-Pichler-Tempsky, Vienna, 1981), pp. 305–8.

—— 'Cloning', *Philosophy*, vol. 57 (1982), pp. 201–9.

Chan, C. K., 'I.Q., heredity, and social achievement', *Science for the people* (1976).

—— 'Lee Kuan Yew and I.Q.', *Aliran Monthly* (May 1984), pp. 19–20.

—— 'I.Q., ideology and social policy', *Aliran Monthly* (June 1984), pp. 7–9.

Chedd, G., 'The making of a gene', *New Scientist*, 30 Sept. 1976.

Cohen, S. N., 'The manipulation of genes', *Scientific American*, vol. 233 (1975), pp. 24–33.

Corea, G. *et al.*, *Man made women: how new reproductive technologies affect women* (London, Hutchinson, 1985).

Cottingham, J. G., 'Warnock and after: moral and legal status of embryos', paper presented at the 1986 Annual Conference of the Association for Legal and Social Philosophy.

Council for Science and Society, *Human procreation: ethical aspects of the new techniques* (Oxford University Press, Oxford, 1984).

Davis, B. D., 'Prospects for genetic intervention in Man', *Science*, vol. 170 (1970), pp. 1279–83.

Dawkins, R., *The selfish gene* (Oxford University Press, Oxford, 1976).

Devlin, P., *The enforcement of morals* (Oxford University Press, Oxford, 1965).

Dubos, R., *Man, medicine and environment* (Penguin, Harmondsworth, 1970).

Dunstan, G. R., 'Warnock reviewed', *Crucible, Quarterly Journal of the Board for Social Responsibility*, (Oct.–Dec. 1984), pp. 148–53.

Dworkin, R., *Taking rights seriously* (Duckworth, London, 1977).

Dyson, A., 'After Warnock: questions to the church', *Crucible, Quarterly Journal of the Board for Social Responsibility*, (Oct.–Dec. 1984), pp. 154–61.

—— Editorial: 'Warnock', *The Modern Churchman*, vol. XXVI, no. 4, 1984, pp. 1–2.

Ebling, F. J. (ed.), *Biology and ethics* (Academic Press, New York, 1969).

Ebon, M., *The cloning of man* (Signet, New York, 1978).

Edwards, R. G., 'Fertilization of human eggs *in vitro*: morals, ethics and the law', *Quarterly Review of Biology*, vol. 49 (1974), pp. 3–26.

—— and Purdy, J. M., *Human conception in vitro* (Academic Press, London, 1982).

Elster, J., 'Sour grapes — utilitarianism and the genesis of wants' in Sen, A. and Williams, B. (eds), *Utilitarianism and beyond* (Cambridge University Press, Cambridge, 1982).

Etzioni, A., 'Sex control, science, and society', *Science*, vol. 61 (1968), pp. 1107–12.

—— *Genetic fix: the next technological revolution* (Harper Colophon, New York, 1975).

Fletcher, J. C., 'Ethical issues in and beyond prospective clinical trials of human gene therapy', *Journal of Medicine and Philosophy* (1985), p. xxx.

Flower, M., 'Gene therapy: proceed with caution', *Hastings Center Report*, vol. 14, no. 2 (1984), pp. 13–17.

Floyd, S. L. and Pomerantz, D., 'Is there a natural right to have children?' in Arthur, J. (ed.), *Morality and moral controversies* (Prentice-Hall, Englewood Cliffs, NJ, 1981), pp. 131–8.

French, M., *Beyond power: women, men and morals* (Jonathan Cape, London, 1985).

Fried, C., *Right and wrong* (Harvard University Press, Cambridge, Mass., 1978).

Friedmann, T. and Roblin, R., 'Gene therapy for human genetic disease?', *Science*, vol. 175 (1972), pp. 949–55.

Galton, F., *Hereditary genius*, 2nd edn (London, 1892).

Glover, J. C. B., *Causing death and saving lives* (Penguin, Harmondsworth, 1977).

—— *What sort of people should there be?* (Penguin, Harmondsworth, 1984).

Goodenough, U. and Levine, R. P., *Genetics* (Holt, Rinehart & Winston, London, 1974).

Goodfield, J., *Playing God: genetic engineering and the manipulation of life* (Hutchinson, London, 1977).

Gore, R., 'The awesome worlds within a cell', *National Geographic* (September, 1976).

Greer, G., *Sex and destiny: the politics of human fertility* (Secker & Warburg, London, 1984).

Halsey, A. H., *Heredity and environment* (Methuen, London, 1977).

Hare, R. M., *Moral thinking: its levels, method and point* (Clarendon Press, Oxford, 1981).

Harris, E., 'Respect for persons' in de George, R. T. (ed.), *Ethics and society: original essays on contemporary moral problems* (Macmillan, London, 1968), pp. 111 – 32.

Harris, H. and Watkins, J. F., 'Hybrid cells derived from mouse and man: artificial heterokaryons of mammalian cells from different species', *Nature*, vol. 205 (1965), p. 640.

Harris, J., *The value of life* (Routledge & Kegan Paul, 1985).

Harsanyi, Z. and Hutton, R., *Genetic prophecy: beyond the double helix* (Granada, London, 1982).

Hart, H. L. A., *Law, liberty and morality* (Oxford University Press, Oxford, 1963).

Holden, C., 'Looking at genes in the workplace', *Science*, vol. 217 (1982), pp. 336 – 7.

Honey, C., 'The ethics of *in vitro* fertilization', *The Modern Churchman*, vol. XXVI, no. 4 (1984), pp. 3 – 12.

'Horizon: Brave new babies?', BBC Television, 15 Nov. 1982.

Hull, R. T., 'Philosophical considerations in the growing potential for human genetic control', *Annals of the New York Academy of Sciences*, vol. 265 (1976), pp. 118 – 26.

Illich, I., *Limits to medicine — medical Nemesis: the expropriation of health* (Marion Boyars, London, 1976).

Jones, H. W., 'The ethics of *in-vitro* fertilization' in Edwards, R. G. and Purdy, J. M., *Human conception in vitro* (Academic Press, London, 1982).

Kalmus, H., *Genetics* (Penguin, Harmondsworth, 1948).

Kass, L. R., 'Babies by means of *in-vitro* fertilization: unethical experiments on the unborn?', *The New England Journal of Medicine*, vol. 285, no. 21 (1971), pp. 1174 – 9.

King, J., 'A science for the people', *New Scientist*, 16 June 1977.

Kingdom, E. F., 'The right to reproduce', paper presented at the 1986 Annual Conference of the Association of Legal and Social Philosophy.

Krimmel, H. T., 'The case against surrogate parenting' in *Hastings Center Report*, vol. 13, no. 5 (1983).

Kuhse, H. and Singer, P., *Should the baby live?: the problem of handicapped infants* (Oxford University Press, Oxford, 1985).

Lafollette, H., 'Licensing parents', *Philosophy and Public Affairs*, vol. 9, no. 2 (1980), pp. 182–97.

Lappé, M., 'Reflections on the cost of doing science', *Annals of the New York Academy of Sciences*, vol. 265 (1976), pp. 102–11.

Leach, G., *The Biocrats* (Penguin, Harmondsworth, 1972).

Le Fanu, J., 'Infertile Debates', *New Statesman*, 23 Nov. 1984, p. 13.

Levin, I., *The boys from Brazil* (Michael Joseph, London, 1976).

Life, *Warnock dissected: a commentary on the report of the Committee of Inquiry into Human Fertilisation and Embryology* (Life, Leamington Spa, 1984).

McAuliffe, K., 'The cell seer', *Omni* 8 (February 1986), pp. 54–98.

McLaren, A., 'Why study early human development?', *New Scientist* (24 April 1986), pp. 49–52.

Maclean, S., 'The right to reproduce' in Campbell, T. *et al.* (eds), *Human rights: from rhetoric to reality* (Blackwell, Oxford, 1986), pp. 99–122.

Manser, A. R., 'The concept of evolution', *Philosophy*, vol. 40 (1965), pp. 18–34.

Marx, J. L., 'Gene transfer in mammalian cells: mediated by chromosomes', *Science*, vol. 197 (1977), pp. 146–8.

Mason, J. K. and McCall Smith, R. A., *Law and medical ethics* (Butterworth, London, 1983).

Maynard Smith, J., *The theory of evolution* (Penguin, Harmondsworth, 1958).

Medawar, P. B., 'The genetic improvement of Man' in *The hope of progress* (Wildwood House, London, 1974).

—— and Medawar, J. S., 'Eugenics' in *The life science: current ideas of biology* (Wildwood House, London, 1977).

Mertens, T. R. (ed.), *Human genetics: readings on the implications of genetic engineering* (John Wiley, New York, 1975).

Milunsky, A. and Annas, G. J. (eds), *Genetics and the law* (Plenum Press, New York, 1976).

—— *Genetics and the law II* (Plenum Press, New York, 1979).

Motulsky, A. G., 'Impact of genetic manipulation on society and medicine', *Science*, vol. 219 (1983), pp. 135–40.

Mount, F., *The subversive family: an alternative history of love and marriage* (Counterpoint, London, 1983).

New Scientist, 'Genetic engineers pick out bad genes in foetus', 2 Nov. 1978.

—— 'Curing genetic diseases before birth', 12 April 1984.

Norton, B., 'A "fashionable fallacy" defended', *New Scientist*, 27 April 1978.

O'Brien, M., *The politics of reproduction* (Routledge & Kegan Paul, London, 1981).

Packard, V., *The people shapers* (Futura, London, 1978).

Passmore, J., *The perfectibility of man* (Duckworth, London, 1970).

—— *Man's responsibility for nature* (Duckworth, London, 1974).

Paterson, D. (ed.), *Genetic engineering* (BBC Publications, London, 1969).

Plato, *Laws*, trans. Trevor J. Saunders (Penguin, Harmondsworth, 1970).

—— *Republic*, in *Platonis opera*, vol. IV (Oxford, Clarendon Press, 1957).

Powledge, T., 'Can genetic screening prevent occupational disease?', *New Scientist*, 2 Sept. 1976, pp. 486–8.

Ramsey, P., 'Shall we "reproduce"? I: The medical ethics of *in vitro fertilization*', *Journal of the American Medical Association*, vol. 220, no. 10 (1972), pp. 1346–50.

—— 'Shall we reproduce? II: Rejoinders and future forecast', in *Journal of the American Medical Association*, vol. 220, no. 11 (1972), pp. 1480–5.

Rawls, J., *A theory of justice* (Oxford University Press, Oxford, 1973).

Rescher, N., *Distributive justice: a constructive critique of the utilitarian theory of distribution* (Bobbs-Merrill, New York, 1966).

Restak, R. M., *Premeditated Man: bioethics and the future of human life* (Penguin, New York, 1977).

Rifkin, J., *Declaration of a heretic* (Routledge & Kegan Paul, Boston, 1985).

—— 'Perils of genetic engineering', *Resurgence*, vol. 109 (1985), pp. 4–7.

Roberts, C., 'Positive eugenics' in Rachels, J. and Tillman, F. A. (eds), *Philosophical issues: a contemporary introduction* (Harper & Row, New York, 1972).

Robertson, J. A., 'Surrogate mothers: not so novel after all', *Hastings Center Report*, vol. 13, no. 5, Oct. 1983, pp. 28–34.

Rorvik, D., *Brave new baby* (New English Library, London, 1978).

—— *In his image* (Sphere Books, Aylesbury, 1978).

Rothschild, Lord, 'Risk', *The Listener*, 30 Nov. 1978, p. 717.

Russ, J., *The female man* (Women's Press, London, 1985).

Russell, B., 'Eugenics' in *Marriage and morals* (Allen & Unwin, London, 1929).

Samuels, A., 'The right to reproduce: a reply to Elizabeth Kingdom', paper presented at the 1986 Annual Conference of the Association for Legal and Social Philosophy.

Sidgwick, H., *The methods of ethics*, 7th edn (Macmillan, London, 1907).

Sikora, R. I., 'Negative utilitarianism: not dead yet', *Mind*, vol. 85 (1976).

Singer, M., 'Scientists and the control of science', *New Scientist*, 16 June 1977.

Singer, P., *Practical ethics* (Cambridge University Press, Cambridge, 1979).

—— *The expanding circle* (Clarendon Press, Oxford, 1981).

—— and Wells, D., *The reproduction revolution* (Oxford University Press, Oxford, 1984).

Sinsheimer, R. L., 'An evolutionary perspective for genetic engineering', *New Scientist*, 20 Jan. 1977, pp. 150–2.

—— 'Genetic engineering: life as a plaything', *Forum* (April 1983), pp. 14–15, 70.

Smart, J. J. C. and Williams, B., *Utilitarianism for and against* (Cambridge University Press, Cambridge, 1973).

Smart, R. N., 'Negative utilitarianism', *Mind*, vol. 67 (1958), pp. 542–3.

Smith, A. T. H., 'Warnock and after: legal and moral issues surrounding embryo experimentation', paper presented at the 1986 Annual Conference of the Association for Legal and Social Philosophy.

Snowden, R. and Mitchell, G. D., *The artificial family: a consideration of artificial insemination by donor* (Allen & Unwin, London, 1981).

Stapledon, O., *Last and first men* (Penguin, Harmondsworth, 1963).

Steinem, G., *Outrageous acts and everyday rebellions* (Fontana, London, 1984).

Stent, G. S. (ed.), *Morality as a biological phenomenon* (Dahlem Conferenzen, 1978).

Strickberger, M. W., *Genetics* (Macmillan, New York, 1968).

Sturtevant, A. H. and Beadle, G. W., *An introduction to genetics* (Dover Publications, New York, 1962).

Thomas, D., *The experience of handicap* (Methuen, London, 1982).

Time Magazine, 'Tinkering', 18 April 1977.

Toon, P. D., 'Defining "disease" — classification must be distinguished from evaluation', *Journal of Medical Ethics*, vol. 7 (1981), pp. 197 – 201.

Tranoy, K. E., 'Asymmetries in ethics: on the structure of a general theory of ethics', *Inquiry* 10 (1967).

Vines, G., 'New tools to treat genetic diseases', *New Scientist*, 13 March 1986, pp. 40 – 2.

Vyvyan, J., *The dark face of science* (Michael Joseph, London, 1971).

Wade, N., 'Gene-splicing: Senate bill draws charges of Lysenkoism', *Science*, vol. 197 (1977), pp. 348 – 59.

Walker, A. D. M., 'Negative utilitarianism', *Mind*, vol. 83 (1974), pp. 424 – 8.

Walters, W. and Singer, P. (eds), *Test tube babies* (Oxford University Press, Melbourne, 1982).

Warnock, M., *A question of life: the Warnock Report on human fertilisation and embryology* (Blackwell, Oxford, 1985).

—— (Chairman), *Report of the Committee of Inquiry into Human Fertilisation and Embryology* (The Warnock Report), Cmnd. 9314 (HMSO, London, 1984).

'Where there's life', Yorkshire Television, 30 July 1986.

Whitehouse, H. L. K., *Towards an understanding of the mechanism of heredity*, 3rd edn (Edward Arnold, London, 1973).

Williams, B., *Problems of the self* (Cambridge University Press, Cambridge, 1973).

Williams Committee, *Report of the Working Party on the Practice of Genetic Manipulation* (HMSO, London, 1976).

Wilson, E. O., *Sociobiology* (Harvard University Press, Cambridge, Mass., 1975).

—— *On human nature* (Harvard University Press, Cambridge, Mass., 1978).

Zamyatin, Y., *We*, trans. Bernard Guilbert Guerney (Penguin, Harmondsworth, 1972).

Newspaper articles

Daily Express, ' "Father" who made sex obsolete', 8 March 1978.

Daily Mail, 'Tragic girl loses her case against being born', 20 Feb. 1982.

—— 'The magic key that unlocks the future', 26 June 1982.

—— 'The men who want to be Mother', 19 May 1986.

Edwards, R., 'Embryonic research: Dr. Edwards replies', *Sunday Times*, 10 Oct. 1982.

Egginton, J., 'US surgeons operate on unborn babies', *Observer*, 22 Nov. 1981.

Guardian, 'Judges rule that handicapped girl cannot seek damages for being allowed to be born', 20 Feb. 1982

———— 'Embryo grown in dish starts Life objection', 11 April 1984.

———— 'Why the embryologists are not acting irresponsibly', 27 Feb. 1985.

———— 'Test tube pioneers condemn Powell bill', 12 March 1985.

———— 'Convention will not aid embryo research foes', 21 March 1985.

———— 'Take genes test, couples urged', 28 Aug. 1985.

———— 'Test tube challenge to the family', 28 Aug. 1985.

———— 'Why you mustn't put all your eggs in one test tube', 16 Oct. 1985.

———— 'The microbes that are undermining the theory of evolution', 28 Oct. 1985.

———— 'Embryo tests await go-ahead', 19 Nov. 1985.

———— 'Cancer team wants human embryo cells', 20 Nov. 1985.

———— 'Tale of an egg that baffled scientists', 21 Nov. 1985.

Kayhan, 'Crucial role for genetic testing', 21 Aug. 1977.

Observer, 'Problem births for the "superbabies"', 2 March 1980.

———— 'Is the family a failure?', 3 Oct. 1982.

———— 'The shape of babies to come', 21 Nov. 1982.

———— 'Lee puts hope in "designer genes"', 18 Sept. 1983.

———— 'Treat the embryo with respect', 16 Feb. 1985.

Perrick, P., 'Feminists can't win', *The Times*, 4 August 1986.

Robertson, G., 'Go forth and multiply but only if the Minister agrees', *Guardian*, 18 March 1985.

Singapore Monitor, 'Getting the right answer for Mum', 25 Jan. 1984.

Singapore *Straits Times*, 'School entry rules changed', 24 Jan. 1984.

Sunday Mirror, 'Threat of the carbon copy', 11 Sept. 1977.

Sunday Times, 'Sex-choice promise', 18 July 1982.

Sunday Times Magazine, 'The proxy fathers: sowing the seeds of despair', 11 April 1982.

The Times, 'The moral dilemmas of the biological revolution', 28 June 1977.

———— 'By-passing a block to conception', 27 Feb. 1978.

———— 'Lock up your laboratories', 31 Aug. 1978.

———— 'California's germinal brains bank', 3 March 1980.

———— 'How shall a child know its parent?', 19 Feb. 1982.

———— ' "Wrongful life" no cause of action', 20 Feb. 1982.

———— 'Nobel no-go, but still a quest for genius', 24 Sept. 1982.

———— 'Boycott human embryo experiments, BMA says', 28 Sept. 1982.

———— 'Refusal by doctor affects thousands', 24 March 1986.

———— 'Genetic engineering: 1 "Risk profile" helps in checking tendency to inherited illnesses', 28 April 1986.

———— 'Genetic engineering: 3 Tomorrow's cures in the making', 30 April 1986.

———— 'Fatherless families foster crime and violence, study finds', 4 Aug. 1986.

Veitch, A., 'Fertile imaginings', *Guardian*, 30 May 1984.
——— 'Genetic "print" that could point the finger', *Guardian*, 11 March 1985.
——— 'Gene probe "can cut haemophilia babies by third" ', *Guardian*, 25 April 1985.
Whitehorn, K., 'A case of Noblesse O'Blige', *Observer*, 9 March 1980.
Winston, R., 'Why we need to experiment', *Observer*, 10 Feb. 1985.
Wright, P., 'Men in the news: Mr. Steptoe and Dr. Edwards', *The Times*, 27 July 1978.

Index